Ethnozoology of Egede's "Most Dreadful Monster," the Foundational Sea Serpent

CONTRIBUTIONS
IN ETHNOBIOLOGY

CONTRIBUTIONS IN ETHNOBIOLOGY

Marsha Quinlan, Justin M. Nolan, and Sarah Walshaw, Series Editors

Contributions in Ethnobiology is a peer-reviewed monograph series presenting original book-length data-rich, state-of-the-art research in ethnobiology. It is the only monograph series devoted expressly to representing the breadth of ethnobiological topics.

Explorations in Ethnobiology: The Legacy of Amadeo Rea
Marsha Quinlan and Dana Lepofsky, Editors

Sprouting Valley: Historical Ethnobotany of the Northern Pomo from Potter Valley, California
James R. Welch

Secwepemc People and Plants: Research Papers in Shuswap Ethnobotany
Marianne B. Ignace, Nancy J. Turner, and Sandra L. Peacock, Editors

Small Things Forgotten: Artifacts of Fishing in the Petén Lakes Region, Guatemala
Prudence M. Rice, Don S. Rice, and Timothy W. Pugh

Ainu Ethnobiology
Dai Williams

Sahnish (Arikara) Ethnobotany
Kelly Kindscher, Loren Yellow Bird, Michael Yellow Bird, and Logan Sutton

Ethnozoology of Egede's "Most Dreadful Monster," the Foundational Sea Serpent
Robert L. France

Ethnozoology of Egede's "Most Dreadful Monster," the Foundational Sea Serpent

Robert L. France

Society of Ethnobiology

2021

"It is reported that the waters about Greenland are infested with monsters."
— *The King's Mirror (Konungs Skuggsja)*, ca. 1280

"One sees monstrous fishes there [Greenland] that are very marvelous."
— Antoine de la Sale, *La Salade,* 1442

"As to the non-descript sea-monsters so often talked of, few creditable persons have had ocular evidence of their existence. But what to think of the following relation given openly by such a worthy man as Mr. Paul Egede, it is difficult to decide."
— David Cranz, *The History of Greenland,* 1767

Photograph of an illustration at the Canadian Fossil Discovery Centre in Manitoba that depicts conditions when the region was covered by the North American Inland Seaway (see Appendix 1.3).

For my childhood friends Mike Cipryk and Jim Cushon, and experiences fondly remembered of hunting for fossils of prehistoric marine creatures and of fleeing Pellucidarian monsters, both of which inhabited our Manitoba neighborhood and imagination. And for my young niece Taylor Smith, who is an enthusiastic fan and frequent visitor to the skeletal remains of "Bruce," the World's largest displayed mosasaur (see Appendix 1.3), and my cousin John Pawling, who loves watching television shows about mysteries filled with the sort of speculative questions that are also prevalent herein.

Library of Congress Control Number: 2021915158

ISBN 978-0-9990759-4-4 (paperback)
ISBN 978-0-9990759-5-1 (PDF)

Society of Ethnobiology
Boston University Archaeology Room 345,
675 Commonwealth Ave., Boston, MA 02215

Cover photo: Completion of the transformation of Egede's "terribly big sea-creature" into a "most dreadful monster" of fantasy. Illustration by James Stewart in Robert Hamilton's *Amphibious Carnivora* (1839), part of William Jardine's "Naturalist's Library" series.

Table of Contents

List of Figures

List of Tables

Acknowledgments

My thanks to the library staff at Dalhousie University for aiding in the acquisition of many sources used in the production of this monograph, and to the reviewers of the present manuscript through the Society of Ethnobiology, including editor Sarah Walshaw. Illustrations are in the public domain or Wikimedia Commons, or obtained through permissions from Hill and Wang (Figures 2.2 and 2.3), Bob Eggleton (Figure 2.5), Penguin Random House LLC (Figure 2.7), and DK Publishing (Figure 5.8). All photographs are by the author unless otherwise specified.

Introduction

In the great heyday of natural history during the nineteenth century (Barber 1980), many of the world's leading scientists believed the study of sea monsters to be a pursuit that was not only legitimate, but also meritorious (e.g., Brown 1990; France 2019a; Lyons 2009; Regal 2012). Harvard professor Jacob Bigelow (1820), for example, referred to the 1817 appearance of the Gloucester Sea Serpent in New England as being "the most interesting problem in the science of natural history." Consequently, dozens of articles about sea serpents/monsters were published by learned professional and amateur naturalists in leading journals, such as *The Zoologist, Nature,* and the *American Journal of Science and Arts* (Lyons 2009; Westrum 1979). As a result, "no marine environmental historian worth his or her salt can afford to ignore early…. [accounts of] sea serpents" (Bolster 2012:91). Today, due to overt confirmation biases and failures to employ parsimony (i.e., Occam's razor) in the proposition of explanations about the identity of mystery animals behind the observed phenomena, the pseudoscience of cryptozoology is generally regarded as a fringe activity (e.g., Dendle 2006; France 2019a; Loxton and Prothero 2015; Rossi 2016). Published peer-reviewed scholarship (in contrast to online blogs, websites, videos, and the like, which abound) on lake monsters and sea serpents has, in consequence, become largely the purview of historians, anthropologists, and other social scientists (Bartholomew 2012; Brink-Roby 2008; Brown 1990; Burns 2014; Cheezum 2007; Papadopoulos and Ruscillo 2002; Regal 2012; Starkey 2017; Stothers 2004). "Largely," but not exclusively, as the efforts of a few biologists, through the benefit of having much greater knowledge and insight about marine biodiversity than their nineteenth-century counterparts, have resulted in known animals being posited for earlier sightings of what had hitherto been unidentified marine objects, or "UMOs" (e.g., Brongersma 1968; France 2019a; Galbreath 2015; Paxton and Holland 2005; Paxton et al. 2005; Woodley et al. 2011, 2012).

Conservation biology needs to consider historical interpretations to ensure comprehensive assessments about present-day conditions and to generate robust proscriptions about future restoration goals (Crumley 1994; France 2016a; Kittinger et al. 2015; Meine 1999). Today's ecosystems are very much shaped by both past actions and also corresponding societal attitudes (e.g., McDonnell and Pickett 1993; Szabo and Hedl 2010). Because the investigation of past conditions often necessitates detective work (McClenanchan 2015), historical ecological research is a forensic pursuit (Alexander et al. 2017). The use of information, compiled from a wide variety of non-standard sources (e.g., Al-Abdulrazzak et al. 2012; Guidetti and

Micheli 2011; McClenachan et al. 2012; Saenz-Arroyo et al. 2005, 2006), has proven helpful in marine studies that address the conundrum posed by the "shifting baseline syndrome" (*sensu* Pauly 1995), wherein each generation redefines what is deemed "natural" according to its temporally truncated view of an increasingly depauperate world. Parsons (2004), in turn, suggested that folkloric tales of sea monsters can provide antecedent information useful for interpreting historic changes in biogeography.

Under what discipline then does the historical forensic study of sea serpents/monsters reside? Concerned with studying the dynamic interrelations between people, biota, and the environment in both past and present times, ethnozoology is part of the overall discipline of ethnobiology (Anderson et al. 2011), and is interdisciplinary, subsuming elements from conservation biology, anthropology, historical ecology, archaeology, environmental history, and psychology, in addition to its foundational roots in zoology and ethnology. Ethnozoology, therefore, focuses on understanding the roles played by animals in human society, both those that are utilitarian as well as those that are cultural, religious, artistic, and philosophical (Posey 2000; Szabo 2008). Sometimes referred to as "folk zoology" (e.g., Forth 2016; Franklin 1990; Hunn 1978), this type of research is often concerned with people's everyday understanding of the biological world in terms of how they comprehend and categorize fauna. Most importantly, ethnozoology entails the ecological, cognitive, and symbolic study of fauna, regardless of whether those animals physically exist or are imaginary (da Silva Vieira et al. 2017). In addition to ethnozoologists and folklorists (e.g., Meurger and Gagnon 1988; Ritvo 1997), serious scholarship about animals which are either imaginary or for whose existence is presently unproven in contemporary Western or International science and culture, has attracted the interest of anthropologists, philosophers, paleobiologists, and historians of science and popular culture (e.g., Asma 2009; Bartholomew 2012; Hackett and Harrington 2018; Hurn 2020; Loxton and Prothero 2015; Naish 2017; Regal 2011; Williams 2015).

It is important to make a distinction. In their seminal 1988 book, *Lake Monster Traditions: A Cross-Cultural Analysis*, Michel Meurger and Claude Gagnon argue that the investigation of undiscovered animals which are presumed to be real belongs to the purview of folklore (an ethnological social science), not of zoology (a biological physical science). The folkloric or ethnozoological study of the occurrence and taxonomy of imaginary (e.g., mythological or spiritual), mystery (e.g., folk or symbolic classifications), or real (e.g., rare or extirpated) animals, spanning history and across different cultures, is a rich field of scholarly activity (e.g., Forth 2020; Jaffe 2013; Stothers 2004; Svanberg 1999), just as is the pursuit of natural history (e.g., Bartholomew 1986; Del-Claro et al. 2013; Peters 1980). Both, however, are distinct from the physical science of zoology or ecology. In contrast, cryptozoology, the contemporary search for hidden animals, despite the insistence of its defenders that it is a legitimate and empirical form of zoology or biogeography, is largely a fringe activity, often conducted by amateurs in a vacuum of defensible methodology (e.g., Das 2009; France 2019a; Hill 2001; Loxton and Prothero 2015).

With literally hundreds of eyewitness accounts of observed UMOs spanning centuries and across all seven seas, the literature on sea serpents/monsters is extensive and diffuse (e.g., Harrison 2001; Hebda 2015; Heuvelmans 1968; O'Neil 1999; Oudemans 2007[1892]; Thomas 2011). As such, it can be useful to zero in on an individual or limited number of accounts from the entire corpus of sightings, with selection based on either their widespread fame or their perceived reliability (e.g., France 2017; Gould 1930). One such encounter, documented by Hans Egede from a sighting that occurred in 1734 off the coast of Greenland, has since become a much-discussed staple of cryptozoological lore. Furthermore, this particular sighting is of historical significance in consequence of occurring, as it did, between the Renaissance *Hic Svnt Dracones* ("here are dragons") phase of Nordic map making (Nigg 2013; Van Duzer 20014) and the nineteenth-century natural history/science period of attempting to understand what such purported "dragons" might actually have been or perhaps still are (Lyons 2009; McGowan-Hartman 2013). Indeed, in Oudemans' (2007[1892]) comprehensive and diachronic compendium of more than 200 sightings of what he termed "the Great Sea-Serpent," he situates the Egede encounter at number "5" in the temporal sequence. For Gould (1930), Egede's sighting is the premier one in his overview of reports he considered to be credible enough to be worthy of detailed examination. It is therefore not a reach to consider that given the degree of prolonged worldwide attention that has ensued, the Egede sighting can actually be regarded as being foundational in terms of initiating the entire modern cultural phenomena of sea serpents. For, as McGowan-Hartman (2013:50) states,

> Sightings of sea serpents go back in the historical record to ancient times; the modern sea serpent, however, is considered by many to have been first reported in 1741 by Hans Egede, a Norwegian pastor conducting missionary work in Greenland. His description would come to fit what by the mid-nineteenth-century was referred to formally as 'The Great Sea Serpent.'

The purpose of the present monograph is, in Part I, after describing the encounter (Chapter 1), to provide, in Chapter 2, an examination of the varying interpretations that have been advanced over almost three centuries to explain the Egede UMO, and to use this as a window into the world of shifting opinions of observers commenting upon such maritime "natural historical anomalies" (*sensu* Paxton and Shine 2016). This is only the second such diachronic ethnozoological study to be so-entitled for a particular sea serpent/monster (see France 2019a). In the present case, the time period considered spans that from the "age of wonder" (*sensu* Holmes 2008) of the eighteenth to early nineteenth centuries, characterized by earnest yet naïve naturalists, through the "age of contradiction and transition" (*sensu* Lyons 2009) of Victorian avocational naturalists and professional natural scientists, to the contemporary period of conflicting and sometimes antagonistic opinions held by mainstream scientists ("eggheads") and passionate amateur cryptozoologists ("crackpots") (*sensu* Regal 2011). The monograph continues, in Part II, with Chapter 3 addressing environmental concerns germane to the Anthropocene through positing a new parsimonious (i.e., non-cryptozoological) explanation for the Egede UMO, one that has implications for understanding the historical baseline of when and how humans have deleteriously influenced marine biodiversity.

Holistic studies of the ethnozoology of aquatic monsters move beyond documenting the social history of the sightings (Bartholomew 2012; Williams 2015), and often assume a scholarly path as sinuous as the elusive beasts themselves, touching upon, as they do so, a number of disciplines (Burns 2014; Cheezum 2007; France 2019a). It is the same here, as the monograph continues in Part II, with scientifically-based Chapter 4 taking a tangential voyage to propose and then examine a new possible candidate for the animal behind the Egede UMO. Then, the monograph, in culturally-focused Part III (Chapter 5), considers the influence through time of the Egede encounter vis-à-vis two famous paleontological gaffes, later sea monster sightings, modern cryptozoology musings, as well as contemporary imagination. The important point is that because aquatic monsters are mental constructs (Magin 1996; Meurger and Gagnon 1988), and are often metaphors (Asma 2009; Dendle 2006; Perry 2016), like all such creatures born of imagination (Mullis 2019), their cultural influence transcends physical science and becomes truly catholic, requiring an ethnozoological lens of examination. And the route conceptually swum by these enigmatic sea creatures, from monstrous origins to enlightened mundanity in the natural world (e.g., Szabo 2008), and then continuing into modern popular culture (Hackett and Harrington 2018; Jylkka 2018; Magin 2016), leaves a large, serpentine wake. The Afterword considers several precedents for the sighting. Finally, because, "men really do need sea-monsters in their personal oceans" (Steinback 1968:28), the monograph concludes, in the Appendix, by presenting a guide identifying locations where one can observe such monsters today. This provides further evidence for the pervasive ethnozoological influence of sea monsters in contemporary culture. In short, this monograph examines how it is that a seemingly innocuous illustration of a sea monster on an eighteenth-century map can be of such lasting import as to find its way into a scholarly book on the North Atlantic colonization by the Norse (Petersen 2000).

PART I

HISTORY AND HISTORIOGRAPHY

CHAPTER 1.

The Eyewitnesses, Their Times, and the Canonical Account in its Various Forms

The Setting: Monster-Infested Waters of the Early to Late Modern North

Olaus Magnus was a Swedish Catholic priest who fled to a life of exile of Italy following his country's conversion to Lutheranism. Unable to assume his duties as titular Archbishop of Uppsala, he instead directed his energies to producing, in 1539, the *Carta marina et descriptio septemtrionalium terrarium ac mirabilium*, which is described as being "the largest, most detailed, and most accurate map of any part of Europe up to that time" (Nigg 2013:10), and of which two extant copies are known to exist (see Appendix 1.7). Composed of nine sheets and measuring 125 by 170 cm, "the map's waters in all directions teem with sea monsters born of mariner's tales, literary tradition, Olaus's observations, and an artist's imagination" (Nigg 2013:96).

Despite their fanciful depiction, it is important to recognize that sea creatures, including monsters, were more than mere decorations filling empty space or as visual metaphors of subsurface dangers. So, in addition to symbolizing God's omnipotence and fear of the unknown, sea monsters on maps were deliberate representations of early-modern Europe's perceptions of nature and understanding of the world (Hoage 2017; Starkey 2017). As such, they were "graphic records in the geography of the marvelous" (Van Druzer 2014:11). In this regard, such maps are better thought of as illustrations of ideas about the world as opposed to being accurate geographic depictions; and it is for this reason that they were acquired as artistic products rather than as navigational aids (Hoage 2017). In particular, cartography and the illustration of sea monsters was closely related to the development of natural history during the Renaissance (Davies 2016).

The creatures shown on the *Carta marina* (Figure 1.1) are therefore meant to be representations of actual marine life observed by fishermen and sailors, and as such, are described in an accompanying detailed key. In 1555, Magnus expanded upon the descriptions of his monsters in the book *Historia degentibus septentrionalibus …*, which was translated into Eng-

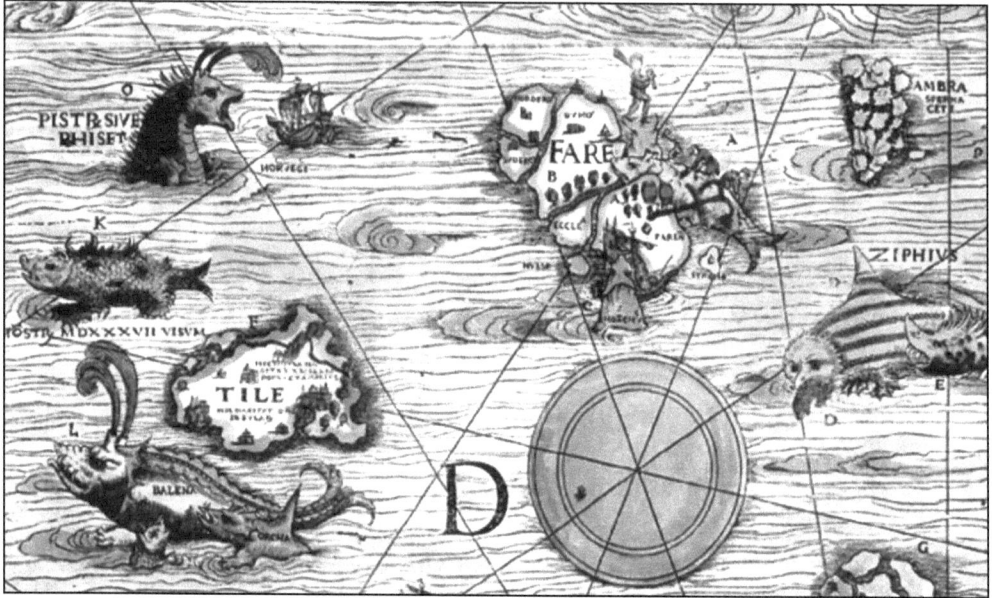

Figure 1.1. A section of Olaus Magnus' 1539 *Carta marina* (in Nigg 2013:13) depicting a shoal of sea monsters surrounding the Faröe Islands and the imaginary Island of (Ultima) Thule.

lish in 1658 (*A Compendious History of the Goths, Swedes, and Vandals and Other Northern Nations*). Not only is the *Carta marina* considered to be the major source of sea monster iconography and lore in the Renaissance, it is situated at the transition between medieval thought and modern inquiry such that its influence on the development of zoological illustration and the nascent science of zoology was profound (Nigg 2013; Van Druzer 2014). For example, Magnus' depictions of sea monsters were copied by Gerardus Mercator for his globe of 1541, by Sebastian Münster for his *Monstra marina et terrestrial* chart of 1544, and most significantly by Conrad Gesner—considered to be the father of modern zoology—for his *Historiae Animalium* natural history of 1551–1558, in addition to many subsequent maps and keyed charts, such as that by Abraham Ortelius (see below).

By the time of the Egede sighting (see below) at the transition between the Early and Late Modern Periods, sea monsters in the European mindset were formulated from the confluence of classical mythology, biblical scripture, medieval bestiaries, and Renaissance natural history and cartography (Jorgensen 2018; Loxton and Prothero 2015; Sweeney 1972). These pan-European mental constructs permeated Scandinavian consciousness to which they conjoined endemic Norse traditions (Meurger and Gagnon 1988; Thomas 2011; Van Duzer 2013), to create folkloric belief in seas that contained all manner of fantastic and fearsome creatures (Szabo 2008). Specifically, back in Magnus' day, the sea and what it harbored were regarded as a source of great wonder, adventure, and surprise (Starkey 2017), as explained in Magnus' chapter on "Monstrous Fishes" in his 1555 book (in Starkey 2017:31):

> The vast Ocean presents a wonderful spectacle to every nation in its swirling waters. It exhibits its various offspring, which strike us not in their wonderful size and similarity to constellations but rather through their threatening shapes, so that there appears to be nothing hidden either in the heavens, or on earth, or in earth's bowels, or even among household tools, which is not preserved in its depths. In this broad expanse of fluid Ocean, receiving the seeds of life with fertile growth, as sublime nature ceaselessly gives birth, an abundance of monsters is found.

And the best place to find mysterious beasts was at northern latitudes (Houwen and Olsen 2001; Jorgensen 2018), as Magnus explains:

> Also, I must add, that on the Coasts of Norway, most frequently both old and new Monsters are seen, chiefly by reason of the inscrutable depth of the Waters. Moreover, in the deep Sea, there are many kinds of fishes, that seldome or never are seen by men (in Nigg 2013:24).

Furthermore, those living in the far north of Norway, as Adam of Bremen believed, "are to this day so superior in the magic arts or incantations that they ... draw great sea monsters to shore with a powerful mumbling of words" (in Szabo 2008:177).

It is no accident that sea monsters were thought to flourish in northern waters, as this followed the European custom (e.g., Sir John Mandeville's highly influential 1371 travelogue) of populating the margins of the known world with wonders (Starkey 2017). Given the perception that monsters thrive in extreme climates, there is a long tradition, extending back to Pliny and Ptolemy, of locating them in the frigid Arctic (Szabo 2008), that *Ultima Thule* place of great wonder wherein days or nights stretched on for months, seas remarkably froze over, and waters teemed with such marvelous creatures as white bears (i.e., polar bears—*Ursus maritimus*), tusked sea elephants (i.e., walruses—*Odobenus rosmarus*), sea unicorns (i.e., narwhals—*Monodon monoceros*), and who really knew what else. Not without reason then could Adam of Bremen write, in the eleventh century, "in that territory live very many other kinds of monsters whom mariners say they have often seen" (in Starkey 2017:40), and the artist(s) responsible for Hereford Cathedral's famous *Mappa mundi* could situate two such creatures in the northern European seas. Such perceptions continued to exist throughout the Renaissance and into modern times, as witness to Scottish historian William Guthrie writing at the end of the eighteenth century that "the fabulous sea-monsters of antiquity are all equaled, if not exceeded by the wonderful animals, which, according to some modern accounts, inhabit the Norwegian sea" (in Nigg 2013:117). Such folkloric belief even continued into the twentieth century (Teit 1918:197):

According to tradition, the sea-serpent was occasionally seen, especially off the coast of Norway. It had its home at the bottom of the sea, and it rarely came to the surface....When travelling on the surface of the sea, the sea-serpent's body stuck out of the water here and there, and its head reared thirty to forty feet above the surface. It had a serpent-like head, large-eyes, and a long mane similar to masses of seaweed.

Animals depicted on early modern maps were situated deliberately, not randomly, their location identifying biogeographical information pertaining to that particular area (Jorgensen 2018). Scandinavian waters were considered to be inhabited by a diversity of strange animals. According to Magnus (in Nigg 2013:134):

[t]here are monstrous fish on the Coasts or Sea of Norway, of unusual Names, though they are reported a kind of Whales, who shew their cruelty at first sight, and make men afraid to see them; and if men look long on them, they will fright and amaze them.

Such beasts also occurred in abundance farther to the west. In his 1590 edition of *Theatrum orbis terrarium*, regarded as being the first true atlas, Ortelius surrounded Iceland with a bevy of sea monsters, many derived from Magnus' earlier map (Figure 1.2). And just as did his predecessor, Ortelius provided a corresponding keyed description for each beast. Ortelius also made use of a mid-thirteenth-century book known as the *Konungs skuggja* (*King's Mirror*) in which a chapter entitled "The Marvels of the Icelandic Seas" describes marine species and natural wonders with such accuracy that many can be identified from the text (Lehn and Schroeder 2003, 2004; Whitaker 1986; see the Afterword).

One of Ortelius' Icelandic sea monsters (Figure 1.3) is of particular interest as it is a precursor and possible inspiration for the highly influential illustration made of the "most dreadful monster" observed by Egede off the coast of Greenland 145 years later (as presented below). This Ortelius sea monster, referred to as the *Staukul,* is reported to be able to stand upright on its tail for extended periods. Today this is recognized as a normal behavior of many whale species called "spy-hopping," whereby animals raise their heads and chests vertically above the water to look about. The *Staukul* was also notable for being dangerous in consequence of its appetite for human flesh (Van Druzer 2014). Nigg (2013) mentions that the Dutch refer to it as *Springval*, its name deriving from its ability to leap clear of the water. Again, this is the behavior recognized today as the normal "breaching" of many species of whale and also of some sharks. Van Druzer (2014) concludes his section on Ortelius' map of Iceland by making the point that it appeared at a transitional time, based as it was on both accurate geographic details and old textual sources about endemic fauna. As he explains, the map contrarily depicts Iceland as both a wild and mysterious place at the edge of the known world, as reflected

Figure 1.2. A section of the map accompanying the 1590 edition of *Theatrum orbis terrarium* (in Nigg 2013:16) by Abraham Ortelius showing a threatening bevy of beasts off the west coast of Iceland.

Figure 1.3. A close-up from the Ortelius map (in Nigg 2013:17) showing a *Staukul* sea monster off the south coast of Iceland.

by the abundance of fierce-looking sea monsters, but also simultaneously as an island that is properly situated within the known geographical sphere of Europe.

Father and son, Hans and Poul Egede (see below), being themselves northern missionaries, would almost certainly have been familiar with the famous hagiography, *The Life of St. Columba,* written in the seventh century by Adomnan, a monk of Iona (Thomas 1988). Therein is described an incident off the western coast of Scotland which resembles the *Staukul* monster, and which reads remarkably similar to the Egede Greenland encounter (and additionally, as will be seen in Chapter 4, to the normal behavior of basking sharks [*Cetorhinus maximus*] which are often seen breaching in the very same region—Speedie 2001):

> While crossing the open sea between Iona and Tiree he [St. Columba] and those with him in the boat saw—look!—a whale of extraordinary size, which rose up like a mountain above the water, its jaws open to show an array of teeth [whale baleen or basking shark gill-rakers?] … turning back in terror … [The men] only just managed to avoid the wash caused by the whale's motion (Thomas in Szabo 2008:52).

Much more so than for Iceland, it was Greenland, in the minds of Renaissance Europeans, that was considered to be at the very edge of the known world. At the time, and even persisting into the nineteenth century, the New World was viewed as a sort of Garden of Eden (Merchant 2003), peopled by innocents, where there was the strongest possibility that antediluvian creatures still roamed or swam about in remote reaches of the wilderness. Such attitudes would be later picked up by eastern seaboard Americans, explaining the easy acceptance of the existence of the famous Gloucester Sea Serpent, in addition to the Presidential charge given to explorers Lewis and Clark to keep a look out for mastodons during their foray into the unknown western interior (France 2019a). As seas became domesticated, the epicenter for preternatural creatures shifted progressively northward and westward (Szabo 2018). The unknown author of the thirteenth-century manuscript, the *King's Mirror*, writes, for example, that "it is reported that the waters about Greenland are infested with monsters" (Whitaker 1986:6). In his 1442 summary of world geography, *La Salade*, de la Sale writes of the many strange "monstrous fishes" to be seen in Greenland waters of the sort brought to life pictorially by Magnus a century later, including giant "lampreys" that "often attach themselves one to another lengthwise and then bind up a vessel in the sea, that, if it is not quickly shaken loose, will soon be lost" (Enterline 2002:178). Dutifully, the *Carta marina* depicts two marine animals confronting each other off the Greenland coast (Figure 1.4), with so little being known about them that they are not even named. In the key, they are described only as being "two colossal sea monsters, one with dreadful teeth, the other with horrible horns and burning gaze—the circumference of its eye is 16 to 20 feet" (Nigg 2013:134). Nigg makes a case that the two monsters represented are a walrus, with tusks pointing the wrong way around, and the kraken or giant squid, indicated by its necklace of tentacles. Regardless of precise identification, the message from Magnus is abundantly clear: Greenland, where one might encounter such fearsome creatures, was a place not to be trifled with. And so the stage was set for one of history's most famous sea serpent sightings: enter the Reverends Hans and Poul Egede, and their companion illustrator, Reverend Bing.

Nigg (2013), in his discussion about Magnus' legacy, makes the link between the archbishop's 1539 map and 1555 book with respect to the type of monster identified therein as the sea serpent, and the famous Egede sighting of 1734. In doing so, he reprints the highly fanciful illustration of those events made at the height of the nineteenth-century craze about sea serpents, the same image that is used for the cover of the present monograph (and examined

Figure 1.4. A section of Magnus' *Carta marina* (in Nigg 2013:13) in which two unnamed monsters of gigantic and terrifying proportions confront each other in the Greenland littoral, whose waters are shown to be perilous due to the presence of driftwood capable of penetrating the hulls of vessels.

below). But in making such a temporal linkage, an incorrect impression is left based on inference of a more direct connection between the two. This is because during the interceding two centuries between Magnus' and Egede's publications, the way Europeans came to regard and cartographically depict the marvels of nature underwent a complete transformation, as Hoage (2017) examines in a chapter in her thesis called "Death to the Sea Monster." Due to the transition from a belief in myth to the birth of modern science, "by the seventeenth century, the sea monster was an outdated fashion in mapmaking" (Hoage 2017:59), leading to their general exclusion and replacement by, ships. This can clearly be seen in the contrast between Magnus's *Carta marina* and the 1675 map of the German physician, naturalist, and near contemporary to Hans and Poul Egede, Friderich Martens (Figure 1.5).

The Eyewitnesses: The Egedes

Hans Egede (1686–1758; Figure 1.6) was a Danish-Norwegian reverend in the Lutheran Church, whose missionary efforts in Greenland led to the reestablishment of the colonial claims of Denmark after three centuries of silence following collapse of the old Norse settle-

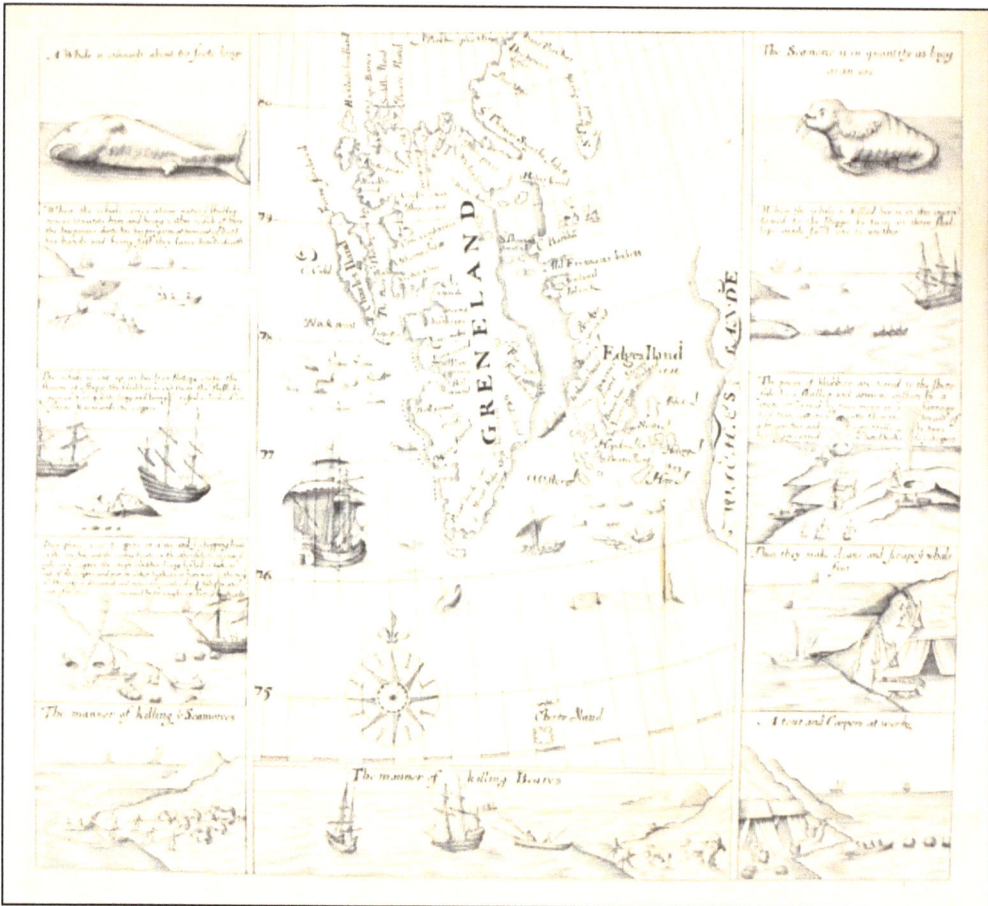

Figure 1.5. Map of Greenland published by Friderich Martens (1675:182) in *Spitzbergische oder Groenlandische Reise-Beschreibung, gethan im Jahre 1671,* which shows that by this time sea monsters have been replaced by ships, and whales are now pictured as mundane (i.e., non-monstrous) animals.

ments there (Garnett 1968). The "Apostle of Greenland," as he would become known, was fascinated by the possibility that there might still be Norse descendants in Greenland. If so, then they would be either active or lapsed Catholics, and therefore, to his Reformation mind, would be in dire need of being brought into the Lutheran version of the True Faith. Empowered with all administrative authority by King Frederick IV of Denmark, Egede set sail in 1721 with his family and about 40 other colonists, many of whom would leave the following year on returning supply ships after suffering through a scurvy-filled winter (Garnett 1968). Over the years that followed, working from his base of Godthab (now the modern-day capital of Nuuk), Egede explored and mapped the Greenland coast in his search for Norse survivors. Meanwhile, he directed his mission in converting the resident Inuit. After returning to Denmark to bury his wife, Egede assumed a variety of ecclesiastical positions of authority, culminating, in 1741, with being made Bishop of Greenland. Today, towns and hotels are named after him, and statues of him stand in both Nuuk and Copenhagen (in

Figure 1.6. The Danish-Norwegian missionaries and naturalists Hans (left) and Poul (right) Egede. Photos courtesy of Wikimedia Commons. See Photo Credits page.

2020, these statues were defaced and calls made for their removal as being symbols of European colonialism). In 1916, the Royal Danish Geographical Society struck a medal in his honor.

Poul Egede (1708–1789; Figure 1.6) took over the Greenland mission for six years when his father was back in Denmark. Later, he too, upon return to Europe, would assume various ecclesiastical positions, including also that of the title of Bishop of Greenland in 1779. Having spent much of his early life in Greenland, Poul, with a Kalaallit woman, produced the first translation of the New Testament into the local Inuit language. Of importance for what follows is the fact that Poul was an accomplished naturalist and one intimately familiar with both the flora and fauna of Greenland, skills which would later result in him being made a fellow of the Royal Norwegian Society of Sciences and Letters.

The Account in Text and Illustrations

On July 6th, 1734, while off the west coast of Greenland, on a voyage from Denmark to the settlement of Disko Bay, an UMO was observed. Seven years later, Hans Egede, who may not have been present on the ship at the time, recounted the sighting as told to him from his son Poul, in his 1741 book, *Grønlands nye Perlustration eller Naturel Historie*. The most accurate English translations from Hans Egede's original Danish are those by Paxton et al. (2005:2) and Thomas (1996:235–236). A combined version of the translation follows, with Thomas' slight discrepancies indicated in brackets:

> Regarding other wonders and monsters of the sea, Tormoder in his *History of Greenland and Iceland* writes about three different kinds, all of which are supposed to have been seen in the waters of Greenland and Iceland. But none of them are in our time [by] us been seen except a terribly big sea-creature [or terribly large sea-animal] which in 1734 was seen in the sea outside the colony at 64 degrees. And was of this form and shape. It was a so enormously big creature [that] its head reached the [ship's] yard arm [or main mast] where the body came out of the water. And the body was as thick as the ship and was 3 to 4 times as long. It had a long pointed nose, and blew [or spouted] like a whale, had big broad flippers, and the body seemed to be covered with a carapace [or scales], and the skin was wrinkled [or uneven] and rough. It was otherwise created at the rear like a serpent and when it went under the water it lifted itself backwards and raised then the tail up from the water a [or one whole] ship's length away from the body.

The subsequent English edition of Hans Egede's book (1745:87–89 in Paxton et al. 2005:2), slightly altered the text. The "terribly big sea-creature" is now referred to as being a "most dreadful monster," the "big broad flippers" are now "great broad paws," and the "carapace" now becomes "shell-work," with the skin being "very rugged and uneven." Another sentence is added that "the under part of its body was shaped like an enormous huge serpent," and now the raising of the tail "aloft" is described as being dependent on the creature "plunging backwards into the sea." Thomas (1996:236) opines this source to be "very poorly translated," with "second rate illustrations."

Other published translations tend to be "woefully garbled" (Gould 1930:13). The 1763 French one states that the UMO was covered in scales and displayed its belly upwards during its plunge back into the water, with the 1763 German translation having the creature actually floating on its back on the surface (Oudemans 2007[1892]). And in 1848, during the heightened global attention about sea monsters brought about by the famous *Daedalus* encounter (Galbreath 2015), the *Illustrated London News* of October 28th, working from a copy of Egede's book in the British Museum, published its own account of the 1734 sighting. Herein (in Oudemans 2007[1892]:97) are repeated the phraseology of "sea-monster," and "its body [being] covered with shell-fish or scales," as well as the new mention made that the tail was a ship-length from the "head" rather than the "body," as had been originally stated. In the 1755 English translation of the widely read and influential 1753 book *Det Förste Forsög paa Norges Naturlige Historie,* by Erik Pontoppidan, Bishop of Bergen (see below), the trait "blew like a whale" becomes "it blew water almost like a whale" (Pontoppidan 1783:180).

The translation sloppiness is somewhat unfortunate since Parish (2020) states that Egede was very deliberate in his choice of wording about the sea-creature based on a concern to facilitate comprehension of what was seen. He does this by providing reference points of familiarity to his readers. The creature's size is compared to that of a ship: its head "reached the yard arm," its body was "as thick as the ship and was 3 to 4 times as long," and its tail was uplifted "a ship's length" distant from the rest of the body. Also the shape and behavior of the creature are compared to things which readers could easily grasp: its rear was a form "like a serpent," and it "blew like a whale."

Poul Egede also published his own first-hand eyewitness account in 1741 (*Continuation af den Grønlandske Mission. Fortater I form a fen Journal fra Anno 1734 til 1740*), which is quite similar to that of his father. Herein, the "very horrible sea-creature" is described as "created below like a serpent," and that it is the act of throwing itself backwards that caused the tail to be raised "thereafter" (Paxton et al. 2005:2). A later recounting by Poul Egede, published in 1789 (*Efterretninger om Grønlandske uddragne a fen Journal fra 1721 til 1788*), concerning the "extraordinarily horrible creature," adds the details that the "breath was not as strong as the whale's," and that the UMO actually breached (i.e., "came out of the water") three times (Paxton et al. 2005:2). Also remarked upon was that the observers were within the proximity of a pistol shot, which, given the range of flintlocks used at the time for dueling (*Wikipedia*), suggests a distance no more than 15 m away. Thomas's (1996) translation continues by mentioning that the evening following the sighting had bad weather, leading, by inference, to the assumption that all had been clear at the time of actual encounter itself. Furthermore, according to Poul, the flippers are described as being "down-hanging" and that the tail "was long" and "then came up" at a distance that was actually "more than" a ship's length away from the body, but that this occurred only after the creature "threw itself backwards." Paxton and colleagues (2005) suggest the 1741 version to be a verbatim transcription of Poul's diary, which was later reworked for the 1789 book.

Poul Egede's 1741 book contains a pictorial map of the west coast of Greenland that includes two representations of the observed UMO (Figure 1.7). The map and illustrations are thought to have been drawn by fellow missionary Reverend Bing, who, from comments made to his brother-in-law Egede and later contained in Bishop Pontoppidan's book, suggests that he had been present on the boat himself and observed the UMO, thereby giving credence, in Gould's (1930) opinion, to the accuracy of the renderings. However, the map's true illustrator is uncertain, as the initials "II.E." seem to imply the hand of the senior reverend, Hans Egede (Loxton and Prothero 2015; Paxton et al. 2005). Acknowledging this, the present monograph will nevertheless go with the established tradition of ascribing the illustration of the UMO to Bing, which, for clarity's sake, has the benefit of reducing the number of references to the pair of Egedes. One drawing depicts the sharp-snouted UMO bounding up from the water, blowing what looks like an exhalation of condensed breath, while its serpentine tail is simultaneously held aloft (Figure 1.8). And the other drawing shows the long and pointed

Robert L. France

Figure 1.7. Map, traditionally but possibly incorrectly attributed to Reverend Bing, showing a portion of the west coast of Greenland (in Paxton et al. 2005:4). Illustration in the bottom-right corner depicts the moment during the famous encounter with the "very horrible sea-creature" when the animal leaps (or breaches), followed later (and inaccurately illustrated as being simultaneous) with the waving of its mysterious, serpentine "tail" in the air. Also shown center-left, above the ship foremast, is a tiny drawing of the airborne "tail" after the animal has re-submerged.

Figure 1.8. Close-up, with a slight enhancement from the original map, of the Egede creature showing its "uneven" or scaled skin, whale-like "spouting" of exhaled vapor (which had held aloft the map title, now cropped out), large pectoral fins or forelimbs, and the possible dorsal fin (in Paxton et al. 2005:4).

tail lingering in the air after the creature has plunged back into the water (Figure 1.9). The nearby ship is shown for scale, in order to indicate that the elevation of the UMO above the water, and also that the distance between the front end of the creature and its tail, are both of comparable size, respectively, to the height of the vessel's yardarm and to its length. The flippers are pointed downward and the surface of the body mottled as per the text, but the creature is seen to be sporting a mouthful of teeth, something not noted in the various de-

Figure 1.9. Close-up from the map of the Egede creature's serpentine "tail" waved aloft (in Paxton et al. 2005:4).

scriptions. Also, halfway down the length of the UMO, a small protuberance is shown jutting out from the body just above the surface of the water.

Although some visual depictions of the Egede encounter would be accurately rendered (e.g., Figure 1.10), others would undergo dramatic reworking. For the article in the *Illustrated London News,* the illustration (Figure 1.11) is altered to match the inaccurate translation in which the rough skin of the UMO has been changed to scales. Pontoppidan's version (Figure 1.12) shows what had been imagined to be the vaporous breath of the creature being transformed into a waterspout that remarkably is sprayed out a full length of the body. The final transformation of the depiction of the UMO from its humble origin as cartographic marginalia to a true sea monster worthy of the phantasmagorical penmanship of a Jules Verne or Edgar Rice Burroughs, occurred with the publication of a plate in Robert Hamilton's volume of Jardine's "Naturalist's Library" (Hamilton 1839). And of course it is this particular representation (Figure 1.13), in which the creature is shown as a "serpentine dragon" (Oudemans 2007[1892]) "prancing like the horse on the boiler of a steam-roller, and spouting a vertical jet of water whose apparent volume is almost as great as that of its whole length" (Gould 1930:13), that has become a staple fixture of coffee-table books (e.g., Sweeney

FIG. 68.—SEA-SERPENT SEEN BY HANS EGEDE, IN 1734, OFF THE SOUTH COAST OF GREENLAND.

Figure 1.10. Accurate rendering of the Egede UMO in Charles Gould's *Mythical Monsters* (1886:32).

Figure 1.11. Altered rendering of the Egede UMO in the 1848 *Illustrated London News* wherein the rough skin has been transformed into true, serpent-like scales (in Oudemans 2007[1892]:99).

Figure 1.12. Altered rendering of Bing's original illustration in Bishop Pontoppidan's highly influential *The Natural History of Norway* (1755:197), showing transformation of original vapor exhalation into a long, fountain-like stream of emitted water.

Figure 1.13. Completion of the transformation of Egede's "terribly big sea-creature" into a "most dreadful monster" of fantasy. Illustration by James Stewart in Robert Hamilton's *Amphibious Carnivora* (1839), part of William Jardine's "Naturalist's Library" series (https://commons.wikimedia.org/wiki/File:Hans_Egede_1734_sea_serpent.jpg).

1972) and online monster websites (e.g., Strange Ark) that are responsible for feeding the hopeful imagination of cryptozoologists ever since (see Chapter 5).

Parish (2020) studied the eighteenth-century natural histories of Greenland produced by Hans Egede in addition to those of David Cranz and Otto Fabricius (see Chapter 2). She makes the point that "the observation and description of the natural history of Greenland was not simply an exercise in recording and categorization, but an exercise in invention, instruction, and interpretation" (Parish 2020:10). All three missionaries write in the context of what each observed as well as about what they had read and heard. Flora and fauna were therefore slotted into taxonomic schema based on a combination of European authoritative texts, the personal experience of the encounter, local Indigenous knowledge, and an anchoring in a pre-existing mindset; in other words, at the "intersection of religion, ethnography, natural history, antiquarianism, and cryptozoology" (Parish 2020:3). In particular, "the monster that he [Egede] recorded in the seas around Greenland can be seen as part of an ongoing process of concatenation of information and evidence from oral and written sources" (Parish 2020:8).

CHAPTER 2.

Chronology and Historiography of Eighteenth to Twenty-First-Century Interpretations

Indigenous Oral Tradition and European Folklore

The previous chapter situated the Egede sighting within the established European tradition of populating the northern oceans with all manner of mysterious and sometimes monstrous sea creatures. Much as his contemporary Gilbert White did in his highly influential *The Natural History and Antiquities of Selborne* (2016 [1789]), Hans Egede, in his own magnum opus, considered cultural aspects of his study region. In particular, he provided many details about the spiritual and mythical beliefs in the oral tradition of the Kalaallit group of Inuit. An important question concerns whether the Egedes' conceptualization of their mysterious sea creature was in any way influenced by the local oral tradition of the Indigenous Greenlanders. Oral traditions in many Aboriginal cultures throughout North America certainly abound with tales of mysterious aquatic beings (Gatschet 1899; Meurger and Gagnon 1988), some of whose serpentine forms depicted throughout the continent (e.g., Davis 1888; Dewdney and Kidd 1973) do in fact resemble what Europeans describe as "sea serpents." Not surprisingly, given the close inter-coupling of Inuit lives with the sea throughout the Arctic Archipelago and their strong shamanism belief system wherein myths were regarded as a legacy of their ancestors, there is a rich panoply of mysterious creatures inhabiting the waters (Boas 1904; Egede 1741, 1745; Nelson 1900; Rink 1875; Wolfson 2001). For example, *Qalupalik* is a human like sea creature who kidnaps misbehaving or wandering children; *Agloolik* is a spirit living under the ice who protects seals; *Tizheruk* is a large snake-like creature with the head of a seal that is said to inhabit Alaskan waters; and the most famous of all Greenland sea creatures is the mermaid-like *Sedna*, regarded by the Kalaallit as the "Sea Woman" or "Mother of Sea Creatures." Labrador and Hudson Bay Inuit believed in the existence of *Nennorluk,* a ferocious and amphibious bear-like creature (but one distinct from polar bears) that consumed seals and, as it walks along the bottom, only becomes visible when entering shallow water (Hynes 2012). Can any of these, or others, be invoked as possibly inspiring the Egede sighting?

A common practice among present-day cryptozoologists—almost all of whom are rooted within Western or International Culture—is to draw attention to antecedent descriptions of monsters from local Indigenous groups or early historical fables as a way to create an impression of zoological verisimilitude for modern contemporary sightings. And so St Columba's use of prayer to banish a water monster from near the head of Loch Ness in 565 CE (Adamnan 1856) is used to bolster sightings of "Nessie" (Costello 1974; Thomas 1988), just as are tales of Native gods and folk heroes battling with dragons or serpents in Maine (Nicolar 1979[1893]; Strong 1998) to do the same for the "Great New England Sea Serpent" (O'Neill 1999), and Labrador Inuit legends likewise for present-day "sea dragons" in Newfoundland (Hynes 2012). Similarly, cryptozoologists often use rock or other artistic renderings of serpentine creatures to make the point that if aquatic mystery animals existed back then in pre-Contact times, it stands to reason that they must also do so today. And so rock art from Maine (Lenik 2010) is used to give credence to European sightings of sea serpents there (O'Neill 1999), just as carvings and rock art from the Pacific Northwest (Hill and Hill 1974; Swords 1991) do the same for sea serpent "Caddy" (LeBlond and Bousfield 1995).

However, and reiterating a previous statement (France 2019a:63):

> It would be wrong to fall into the cryptozoological trap in implying that Aboriginal renderings can be related to later European sightings. Amerindian mythology is distinct from European imagination (Meurger and Gagnon 1988). In other words, serpents seen in shamanic visions are different from those thought to be glimpsed in the water.

Indeed, Meurger and Gagnon (1988) argue that the whole proposition of cryptozoology in transforming such fables into zoological speculations is erroneous as it is based on misguided naturalistic naivety. Folklorist belief, they insist, should not be so naturalized lest it creates a trap of illusory facticity. Aquatic "cryptids" as envisioned by present-day cryptozoologists may be "real" only in the sense that they too, like folkloric animals, are mental constructs that exist within an imaginary landscape (Meurger and Gagnon 1988). For these reasons, Meurger and Gagnon (1988) believed that cryptozoology is better considered from a perspective of the Western traditions of folklore and natural history rather than being conceived of fitting into systems of either local knowledge or mythology of traditional societies. Indeed, Loxton and Prothero (2015) recount the embarrassing saga of confusions arising from cryptozoologists misinterpreting legends and being given distorted information concerning actual events in relation to the most famous aquatic cryptid from Africa. Ethnozoological or folk-biological investigations (Berlin 1992; Medin and Atran 1999) of "ethno-known animals" should therefore not be undertaken in a cavalier fashion (Arment 2004). For the present Greenland case, the question then becomes as to whether animals, be they imaginary from either Indigenous oral tradition or as conceived of by Western folklore, or actual locally ethno-known animals that are presently unrecognized or classified differently by modern zoology, can provide insight about environmental circumstances in (sub)arctic waters?

Van Londen (1996) considered that some myths of marine creatures among Greenland Inuit may represent general coping mechanisms during times of ecological crisis. She cautioned, however, that no clear relationship exists between myth and empirical reality such that "myths operate in a zone of uncertainty, [and] the image that is formed on the basis of reality is not a reflection but a representation" (Van Londen 1996:28). As Meurger and Gagnon (1988) believed, folkloric animals in myths may deviate substantially from reality by forming a world of their own. But what about that hazy borderland lying between the oral tradition of imagined creatures and those customarily encountered by Inuit in their everyday lives? It seems that early European ethnographers were of two minds about this. The eighteenth-century Danish missionary and naturalist Otto Fabricius, who purposely left the mission to live on the land with the native Greenlanders, and whose published works (see next section) are regarded as being among the most accurate and comprehensive faunistic studies ever undertaken (Kapel 2005), championed the local knowledge he received from his Kalaallit hosts. Again and again, in his *Fauna Grøenlandica* (1780), he waxes in amazement about their detailed zoological knowledge. Not so, nineteenth-century Scottish naturalist and Greenland explorer Robert Brown (see next section), who criticizes Fabricius for confusing fabulous and factual animals, thereby leading him to formally describe two new species based on incorrect information received from locals, but which, however, were subsequently erased from the zoological record. Expressing an opinion that was certainly not unique for the Victorian mindset of the colonial era, Brown definitively states that,

> Greenlanders cannot be relied upon for the names of animals … They are not the excellent cetologists we have always been led to suppose, confounding as they do several animals under one name (Brown 1868:360).

Alternatively, Parsons (2004:79) considered that

> although descriptions of sea monsters in [local] folklore may initially appear fanciful, on closer inspection they can sometimes reveal characteristics and features that may be recognizable by marine biologists as diagnostic of living marine animals.

Because of this, folkloric descriptions of mysterious sea creatures contain "seeds of truth" (Parsons 2004:73) about species diversity and distribution in earlier times. In this light, the thesis developed in the next chapter is based on the contention that folklore can inform fishery science, just as cryptozoology can conservation biology (France 2020a), if approached through an eth-nozoological lens (*sensu* France 2019a, 2019c). But what about Egede and the original sighting?

The only Inuit-imagined sea creature with a body form resembling that of the Egedes' UMO is known from the Western Arctic, far removed from Greenland, and therefore does not figure into the local oral tradition. Parish (2020) believes that all three eighteenth-century natural histories of Greenland by Egede, Cranz, and Fabricius reflect the input of Indigenous people as mediated through the missionaries. In contrast, Thomas (2011:11) argues that fa-miliar as he was about Kalaallit perceptions of the marine world, Hans Egede, the ethnog-rapher, does not link, as Hans Egede the naturalist, local knowledge to the "real" mysterious creature that was observed by his son.

> Strangely enough, although Poul and Hans interviewed the locals on many occasions about their customs and information about plants and animals, they apparently never thought to ask them about this strange creature. It might have given us valuable information. Far too often, researchers and others tend to forget that indigenous people are the best possible source of information about local animals (Thomas 2011:11).

Not so it seems in this case, however. It is only a modern cryptozoologist hundreds of years later, Swords (1991), who, explicitly disregarding Meurger and Gagnon's cautions, conjoins fantastical animals from folklore with the real creature seen by Egede, whatever it might have been. As such, it can be safely assumed that for Egede, the "terribly big sea creature" that was observed lay within the northern European tradition rooted in the maps of Magnus and Orte-lius (Nigg 2013) and the text of the *King's Mirror* (Whitaker 1985), as described in the previous chapter and later in the Afterword. And so it is through this Western cultural lens, from early natural history to present-day cryptozoology, that the Egede sighting will subsequently be ex-amined. In support of this approach is the interesting observation made by Brown (1868) that, when it comes to mysterious sea creatures, the cross-cultural influence occurred in the op-posite direction; i.e., due to a close association over centuries, extending back to the medieval European settlements, with inter-marriage and a shared lexicon of certain words (Markham 2015) and no doubt ideas, it was the Inuit who actually adopted myths of fearful sea monsters, such as the Kraken, from the Norse, rather than the other way around.

An extended abstract of Hans Egede's 1741 book was published in the *Royal Society Phil-osophical Transactions* in January 1743: "A Description of Old and New Greenland, or a Natural History of Old Greenland's Situation, Air, Habitude, and Circumstances." Authored by John Green, a medical doctor and secretary of the Gentleman's Society at Spalding, this nine-page-long translation was the introduction to the English-speaking world of Egede's natural history observations. Due diligence is given in the overview to Egede's text about local archaeology, pedology, climate, anthropology, history, and botany. However, when it comes to distilling the Reverend's observations about zoology, the entire encounter with the UMO is absent, airbrushed away into oblivion. This is a glimpse into a time when naturalists were uninterested in dubious sounding stories of mysterious/mythical sea creatures that were best left in place where they belonged on those old Renaissance maps. All this would change, however, following the publication a decade later of Pontoppidan's seminal book (see below).

Fast-forward to today and history professor Christopher Heuer's *Into the White: The Renaissance Arctic and the End of the Image,* which focuses on "the visual poetics of the Far North" (Heuer 2019:18). Although the book begins with mention that "the lands of the Arctic bred terror, worry, and fright, but also the physical and imaginative possibility of new and alternative realities" (Heuer 2019:15), once again no mention is made of sea monsters at all. Even a full chapter dealing with Olaus Magnus, which aptly describes his *Historia* as a "sprawling jumble of storytelling, legend, and natural history" (Heuer 2019:125), does not comment upon the significance of the work and its associated lavishly illustrated map as contributing to the development of natural history. No wonder then that the Egede encounter with the UMO is ignored. The possible mitigating defense that the 1734 Greenland sighting occurred outside the temporal bounds of what technically constitutes the Renaissance, loses weight when one recognizes that the period's bordering dates varied across Europe more or less in relation to distance from Italy. And there was nowhere in greater "Europe" that was further afield from Rome or Florence than Greenland. Heuer's book deals with many more recent objects of study, such as seventeenth-century natural history etchings, nineteenth-century paintings, and even twentieth-century art and architecture, but contains nothing about the critical role that sea monsters have played in all these venues of imagination. The present monograph, in particular, Chapter 5, is positioned as a counterpoint to address this failure to consider the cultural significance of Bing's influential drawing.

Between these two temporal bookends of silence, conjectures have frequently been made with respect to the possible identity of the mysterious marine animal seen by Egede off the coast of Greenland. Over the interlaying two-and-a-half centuries, many would hypothesize about the cryptic nature of the creature, with some opining about the "cryptid" nature of creature, an altogether different thing (as discussed below). This chapter presents an overview of 36 such considerations about the Egede UMO. As is the standard practice when describing sea monster/serpent sightings (Harrison 2001; Hebda 2015; Heuvelmans 1968; O'Neill 1999; Oudemans 2007[1892]), these are presented diachronically. I have, however, aggregated the interpretations into potentially different schools of thought and experience in relation to the occupational and vocational categories of amateur natural historians, professional natural scientists, cryptozoologists, biologists, and wordsmiths. Also, a division is made that separates out and therefore highlights the most informative and important interpretations (i.e., those which are "paramount" and the most influential) from the others (those considered as being "supplementary" in terms of the influence of their hypotheses).

Finally, monographs, such as the present *Contributions in Ethnobiology,* which are published in a format resembling a special issue of a scholarly journal, are ideologically situated between traditional articles (which advance scholarship through iterative development) and typical books (whose task it is to be more comprehensive and inclusive). Such monographs should therefore assume a middle-of-the-road position in terms of balance between containing detailed background material or its referral to in other previously published sources. Therefore, for the present publication, I provide a brief overview of the intellectual context for the different vocational categories used in this chapter (revisited again in Chapter 5) by distilling the lengthy discussions about these foundational themes from my recently published book (France 2019a), and supported therein through the inclusion of numerous references.

Paramount Expositions

Early Natural History

As elaborated on in France (2019a), natural history during the eighteenth and nineteenth centuries blended proto-scientific realism, a sense of wonder, and a belief in the divine mystery of Creation (Barber 1980; Berger 1983; Holmes 2008). The avocational study of natural history became an obsession among the time-rich gentry who, freed from the need to establish a career, run their estates, perform housework, or raise their own children, threw themselves into studying the manifest marvels of God's gift of nature. Natural history societies and magazines abounded, lyceums offered lecture series, curiosity museums flourished, seaside outings became *de rigueur,* and no manor house was complete without a fern-case, butterfly cabinet, shell collection, or aquarium displayed in its parlor (Barber 1980; Knoepflmacher and Tennyson 1977; Mason 2017). This was a period, before biology developed into a formalized profession, when amateurs could make valid contributions to documenting biological diversity and nature's workings, as for example, Henry David Thoreau's studies of plants and species interactions in the woods around Concord, Massachusetts (Berger 2000; McGregor 1997; Walls 1995). Occasionally the efforts of amateurs could even transform understanding of the natural world, as for example, Mary Anning's collection of fossils from the Dorset cliffs of Lyme Regis which contributed to the Victorian understanding of "deep time" and evolution (Emling 2009; Fuller 2001).

Much of the efforts of natural historians were based on a fascination about unusual forms of life and a quest to identify new species. And there was nothing in this regard that was more alluring than the dream of discovering, describing, and documenting the elusive Sea Serpent. This, and the contemporaneous development of paleontology (see below), led to an enormous interest among the general public in what were then (Hawkins 1840), and sometimes still are (Ellis 2003), referred to as "sea dragons," both those species from the distant past and those that might still exist in the present.

For individuals whom had to work for a living, amateur natural historians were mostly assembled from the professions of physicians, apothecaries, teachers, merchants, and ministers (Evans 1993; Stearns 1970). Of these, clergymen were a significant force in the development of natural history (e.g., Paley 2012 [1802]; White 2016 [1789]). Interestingly, in the present context, even the most casual reading of scholarly compilations of sea monster/serpent accounts reveals the predominant role played by reverends, pastors, and ministers in creating the phenomenon. Heuvelmans (1968), for example, entitles his chapter about sixteenth- to eighteenth-century sightings: "The Luckless Bishops." Olaus Magnus and the two Egedes were merely the first in a long tradition in this regard. Part of this might be due to clergymen being well-educated, naturally inquisitive about esoteric matters, and blessed with the luxury of considerable unstructured time (France 2019c). It is no surprise then that the very concept of "sea monsters/serpents" can be said to have had its origin in the natural history writings of one such religious leader.

Erik Pontoppidan was the bishop of Bergen and author of the 1753 classic, *The Natural History of Norway,* which, in addition to considering known fauna, included tales from folklore as well as eyewitness descriptions of encounters with fabulous creatures seen in northern waters. The book became highly influential following publication of its English translation in 1755, and would serve as the standard reference until late in the nineteenth century; i.e., many a sighting of an UMO from around the world during the 1800s would be specifically likened to those initially described by Pontoppidan (e.g., Macrae and Twopeny 1873; see France 2019c). Indeed, more than anyone, it is Reverend Pontoppidan whom can be said to have singlehandedly invented the modern concept of what he termed "sea-serpents," writing: "Like all who are enemies to credulousness I too doubted of the existence of the sea-serpent, when at last my doubt was dispelled by incontestable proofs" (Pontoppidan 1755:196). In his examination of UMOs, Gould (1930:11) believed that any work on sea serpents "which omitted some mention of Pontoppidan (who is more closely associated with that creature than any other man who ever lived) would indeed be like Hamlet without the Prince."

Pontoppidan's main focus and importance with respect to UMOs was twofold: first, to collate and document the rich history of sightings of the Scandinavian *Sæ-orm* (i.e., sea-orm or sea worm), and thereby to rescue it from its former status as a nightmare entity; and second, to give credence to legends concerning the kraken, a beast whose existence would be confirmed in the next century with the discovery of the giant squid (*Architeuthis spp.*). Germane to the present investigation is that, although Pontoppidan does make a clear distinction between the Norwegian sea-orm and the Egede creature observed in Greenland, he labels them both as sea serpents, writing "that this [Norwegian] animal spouts like a whale through its nostrils, as Mr. Egede saw [for his Greenland animal], has never been seen by anybody" (Pontoppidan 1755:199). This suggests, as Oudemans (2007 [1892]) notes, that the Bergen bishop might have been using the inaccurate translation wherein it is water rather than condensed air that was seen being expelled. This is confirmed by Pontoppidan's use of the altered drawing (Figure 1.11). Pontoppidan continues his comparison, noting that,

> the common sea-serpents of our [i.e., Norway's] shore differ from those of the Greenland-coasts, seen by Egede, in having no rough and hard skin, but a smooth one like a mirror, except on the neck, on which it has a mane, resembling sea-weed (Pontoppidan 1755:200).

In the end, it is Pontoppidan who is responsible for taking what Egede had referred to as simply being "a sea-creature," albeit one that was "terribly big," and transforming it into a "sea-serpent," though of a different variety from those endemic and "common" to Norwegian waters.

Victorian Natural Science

As described in France (2019a), the gradual development of the formalized profession of natural science during the nineteenth century underwent a transition from individuals being observers, then collectors, then explainers, then systemacists, to eventually them becoming experimenters, more diligent examiners, and finally bona fide scientific researchers (see also Arment 2004 and Elman 1977). Nevertheless, observations and the acquisition of zoological specimens, even by professionally trained natural scientists, frequently arose from happenstance encounters rather than through systematic surveys made of native fauna. The profession of natural science at this stage had little to do with studying the inner workings of nature but rather was preoccupied with the discovery, description, and formal classification of the planet's biological diversity (Evans 1993; Stearns 1970). The scramble to discover a species that was completely new to science, and thereby be given the honor of having one's own surname ascribed to the animal's Latinized Linnaean binomial title, became an all-consuming passion. The competition was intense, steeped as it was in fierce professional rivalry (Jaffe 2000; see Chapter 5), and national jingoism based on Old- versus New World prejudices (Barber 1980). These same issues were very much at the fore with respect to the question of sea serpents/monsters (Brown 1990). For natural scientists, there was no more prestigious holy grail in this regard than to become the fortunate one able to finally and formally describe that most beguiling of all undocumented creatures: the Sea Serpent.

It is certainly no coincidence that interest in and sightings of "sea dragons" peaked at a time coincident with the popularization and professionalization of paleontology (Loxton and Prothero 2015; Paxton and Naish 2019). Sea monsters/serpents also played a complex role in the early discourse on evolution, with some UMOs that appeared to be hybridizations of recognized animals challenging the very Linnaean classification scheme itself, while simultaneously raising questions about the credibility of expert authority (Lyons 2009; Ritvo 1997).

The result of all this was that, during the nineteenth century, it became almost *de rigueur* for natural scientists, including even the most prominent names in the field, such as Louis Agassiz and Sir Richard Owen (Loxton and Prothero 2015; Regal 2012), to opine about the existence and biological nature of sea monsters/serpents in the pages of prestigious journals (Lyons 2009; Westrum 1979). This, in turn, led to a flurry of contemporaneous books on the topic, of which Henry Lee's would become markedly influential.

Henry Lee was the Curator of the Brighton Aquarium and editor of *Land and Water*, an influential natural history magazine of the time. His 1883 book, *Sea Monsters Unmasked*, produced in association with the International Fisheries Exhibition held in London, is historically significant in being a notable early attempt by someone within the professional biology community to express skepticism about purported sea serpents. In contrast, Lee believed them to be merely misidentified known animals. Today it is recognized that the work suffers from scientific faddism in consequence of the author's over-enthusiastic belief that the giant squid, whose remains were being found washed ashore around the world at the time, can be invoked unilaterally to explain almost all sea serpent sightings.

Lee dismisses Pontoppidan's book as being "full of wild improbabilities and old superstitions" (1883:110). Although commenting on the "high character" of Hans Egede, Lee cautions against a literal acceptance of the illustration due to Reverend Bing's unacquaintance with the animal seen as well as the short period of observation. Lee considers the illustration to be an "impression left on his [Bing's] mind [rather] than the thing [i.e., the creature] itself" (1883:117). With no evidence, he states that Bing invested the drawing with "a character that did not belong to it" (Lee 1883:117). But not enough, however, to disguise the fact that the creature observed was "one of the great calamaries which have since been occasionally met with, but which have only been believed in and recognized within the last few years" (Lee 1883:117). Lee's evidence for such an assertion comes from correctly noting that Egede did not refer to the creature as a sea serpent, as well as his own expert interpretation of the illustration. Working from the inaccurate translation and corresponding illustration noted previously, he proposes that what Egede considered to have been the creature's head was the mantle of a giant squid which, in propelling itself to the surface, thrust "this portion of its body out of the water to a considerable height" (Lee 1883:118), something he mentions as often having himself witnessed. The "supposed tail" appearing some distance from the rest of the body is, in Lee's reckoning, formed by one of the shorter arms of the squid, the suckers being mistaken for scales. According to Lee, the text and illustration incorrectly describe/ show the spout of water coming from the creature's mouth, whereas it really arose from the funnel through which squids are known to expel water. He dismisses the inaccurate artistic embellishments of Pontoppidan's version of Bing's illustration which "deprive it of its original force and character, and of the honestly drawn points which furnish proof of its identify" (Lee 1883:119). In its place, he offers his own version (Figure 2.1) of "the animal which Egede probably saw" (Lee 1883:119).

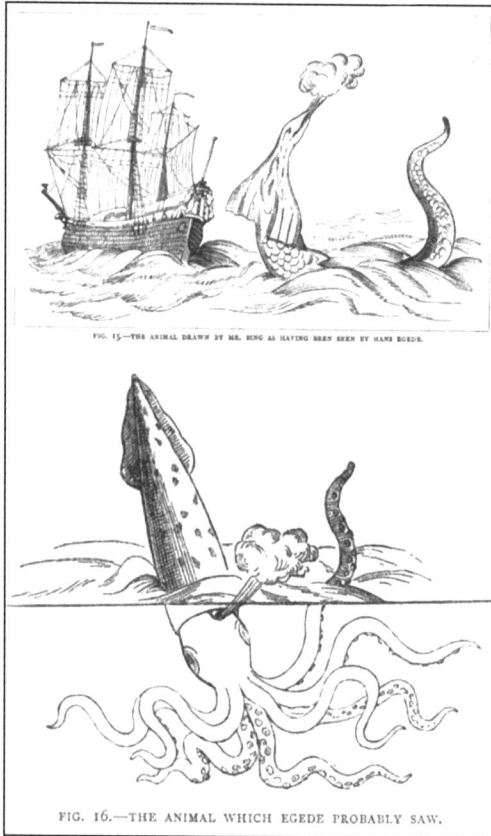

FIG. 15.—THE ANIMAL DRAWN BY MR. BING AS HAVING BEEN SEEN BY HANS EGEDE.

FIG. 16.—THE ANIMAL WHICH EGEDE PROBABLY SAW.

Figure 2.1. The highly influential theory advanced by Henry Lee (1883:120) contrasting what the Egedes thought was seen, and the giant squid that he believes they "probably saw."

Cryptozoology

As detailed in France (2019a), devout Victorians were unable to countenance a world in which prehistoric creatures did *not* swim about, hitherto undiscovered, in remote reaches of the oceans. The reason was that it was simply inconceivable to imagine that God would have created species which would have disappeared before humans were around. Modern-day cryptozoologists, likewise, have a strong belief in the existence of animals that remain hidden from and therefore undescribed by traditional biology. Consequently, "cryptids" are ethno-known animals for which definitive evidence is presently lacking (Arment 2005).

Because cryptozoology operates on: a) a belief that circumstantial evidence is a suitable criterion of legitimacy in support of its claims (Heuvelmans 1988), b) fails to adhere to the scientific protocol of falsifiability (such as championed by Peters 1980), and c) often ignores parsimony or Occam's razor in its praxis (Das 2009), it is frequently regarded with derision as a sham- or pseudo-science (Hill 2011; Schembri 2011; Shermer 2010). Cryptozoology, then, is the study of imaginary animals and manufactured legends by amateur enthusiasts who are largely ignorant of the rules of science and frequently have no relevant training in in the basics of ecology, paleontology, or field biology (Loxton and Prothero 2015). Practitioners err in failing to realize that most aquatic cryptids are real only in the sense that they are mental constructs that often have their origin in folklore, not zoology (Jaffe 2013; Meurger and Gagnon 1988; Naish 2017).

Despite its inherent weaknesses and consequent unsavory reputation (Rossi 2016), cryptozoology is of significance, according to Dendle (2006), from a psychological rather than a biological perspective. In a nod to nineteenth-century nostalgia, it allows a romantic belief in a sense of wonder and surprise at a time when the world has been fully charted and explored; and it provides a vehicle for challenging and even displaying defiance against a culture that unquestionably accepts the certainties purveyed by a select priesthood of arrogant know-it-all scientists (see also Regal 2011). And there are no giants in the field, larger in both reputation and influence, than Antoon Oudemans and Bernard Heuvelmans.

That the Dutch entomologist Antoon Oudemans, an authority on miniscule terrestrial mites, would go on to become the leading proponent for the existence of giant sea serpents, is an irony not lost on many skeptics. That notwithstanding, his compilation of 200 accounts of UMO sightings spanning the globe and over a period of centuries, which was published in 1892 as *The Great Sea-Serpent*, remains the singular work of early scholarship in the field, even if his efforts to posit theories for the various creatures observed are justifiably ill-regarded by today's biologists.

Oudemans provides a detailed review of the variegated publication history of the textual and illustrative versions of the Egede encounter, something which should have, but unfortunately did not, serve as a caution about interpreting the sighting depending on which particular versions were being consulted as sources. In consequence, he is critical of Pontoppidan's conclusions and is particularly and rightly so in regard to the greatly exaggerated illustration appearing in Hamilton's book on natural history. Concerning Lee's illustration of the giant squid as explanation for the Egede creature, Oudemans (2007 [1892]:100) comments : "Well! It looks convincing enough, and there is a savour of ingenious acuteness of wit to it, that might lull the suspicions of a doubting zoologist! What more could be required?" He goes on to state that the "whole fabric [of Lee's thesis] falls to pieces" in consequence of his ideas being "far fetched and thereby impossible" (Oudemans 2007[1892]:100). In support for this conclusion, he notes that: a) when squid are propelled to the surface, their tentacles are stretched downwards, not elevated high above the waters; b) that no squid is physically able to elevate its mantle above the water to such a height; and c) that the expulsion of water from squids' funnels actually occur in a downward direction, contrary to what is shown in Lee's illustration.

Oudemans's own interpretation, however, is as unsatisfying as it is succinct. Fixated on his own pet theory that all sea serpents are long-necked paleo-seals erroneously thought to be extinct, he will not countenance any other explanation: "Pontoppidan is convinced, when seeing Bing's figure, that there are several species of sea-serpents, all belonging to the same genus. I do not wish to discuss this point" (Oudemans 2007[1892]:102). Digging deeper, of especial concern to Oudemans are the illustrations in Pontoppidan's book and the 1848 *Illustrated London News* which change the rough skin of Egede's UMO into scales:

> The body *seemed* to be covered with a hard skin. For truth's sake Egede wrote *seemed*, which is well done; for a hard skin or crust would not have been *wrinkled* when the animal bends its body. Like all known air-breathing sea-animals of those dimensions the animal must of course have under its skin a relatively thick layer of bacon, and I myself have often seen the skin of sea-lions and seals wrinkled, when the animal bent its body in such

a manner as the Sea-Serpent of Egede did. And we shall afterwards repeatedly see that the sea-serpent has no scales but a smooth skin, as seals have
(Oudemans 2007[1892]:100, emphasis in original).

Bernard Huevelmans, widely regarded as the founding-father of modern cryptozoology,
began his career as a zoologist studying mammalian anatomy. In the 1968 English translation of his classic, *In the Wake of the Sea-Serpent,* he reviews over 600 eyewitness accounts
and constructs a taxonomy of nine typologies, of which several types of marine mammals,
some being erroneously (in his mind) thought to be extinct, are proposed as the most likely
candidates for UMOs observed around the world.

Heuvelmans dismisses Lee's giant squid candidacy as being influenced by zoological fashion at the time, and the idea that a cephalopod can leap out of the water as being "farcical."
He slips into sarcasm when considering the reasons why Lee might suppose a giant squid to
swim on the surface lifting one of its tentacles up in the air:

> (1) to see which way the wind is blowing; (2) to wave to one of its fellows;
> (3) to see if it is raining; (4) to give a traffic signal when about to turn;
> (5) to make innocent sailors think that it is a different kind of serpentine
> sea-monster, and incidentally a much less terrifying one (Heuvelmans
> 1968:309).

He goes on to state that despite the title of his book, Lee has not really unmasked anything.
"I suppose the title of his book was too attractive to be dropped," he concludes (Heuvelmans
1968:309).

As for his own interpretation, Heuvelmans begins by praising the observant mind of Hans
Egede with respect to natural history, whose descriptions of which are always "meticulous,
sober and rather dry" (1968:100). Admit-
ting that the English translation is "not entirely accurate," Heuvelmans produces his
own facsimile of Bing's original illustration
(Figure 2.2), that is itself not entirely de-
void of inaccuracies (see below). Several interpretations about the nature of the UMO

20. Hans Egede's 'most dreadful Monster', after Pastor Bing

Figure 2.2. Heuvelmans' rendering of the Bing
illustration in his classic *In the Wake of the Sea-
Serpents* (1968:101), one of the keystone books of
modern cryptozoology. Both the rough skin and
exhalation are shown as per Egede's description,
but the dorsal fin has now been deemphasized by
being removed or hidden beneath the waves.

are advanced. Firstly, based on the size of the ship depicted in the map, the creature can be judged to "have been about 100 feet" in length (Heuvelmans 1968:100). Secondly, as only the tail of "Egede's monster" was serpentine in form, the animal was obviously neither the sea-orm shown by Magnus, nor the Norwegian UMOs noted by Pontoppidan. In consequence, Heuvelmans concludes the creature to have been some other type of marine animal that is unrelated to what is commonly regarded as a "sea serpent." Thirdly, that the "Greenland monster" displayed finned feet or flippers means that it must have been a mammal, which, in a huge leap of supposition, he conjectures could have been amphibious and capable of venturing ashore.

Heuvelmans quotes an account from Pontoppidan which he feels is similar to the Egede sighting, both being sea monsters that are not true sea serpents per se:

> This was not unlike a Sea-calf [i.e., seal] as to the fore-part, and had furred skin. The body was broad and big as a vessel of 50 lasts burthen [i.e., 121,133 L]; and the tail, which seemed to be about six fathoms long [i.e., 12 m], was quite small and pointed at the end (Heuvelmans 1968:106).

This leads him to the following conclusion:

> This is remarkably like 'that most dreadful Monster' described by Egede. The seal's head is very like the pointed snout drawn by Bing. The furred skin could have looked 'rugged and uneven' to the missionaries. The long thin tail and the same—if enormous—dimensions are reported by each. The bishop's beast must have been a mammal and thus had finned feet or flippers, what Egede called 'great broad Paws' (Heuvelmans 1968:106).

Heuvelmans then goes on to liken the Egede encounter to a dozen other sightings of Norwegian UMOs and to place them in his category of "super–otter" with the following traits: slender medium-length neck and long tapering tail; several to half-a-dozen vertical bends in the body; and uniform light grey or beige color. He next describes not only aspects of the super-otter's anatomy based on developing a composite description from the sightings (a valid endeavor), but also, in the kind of cryptozoological flourish of fantasy that invites the ridicule of so many, he goes on to invent aspects of the species' behavior, habitat, and distribution. In his classification scheme, Heuvelmans gives the species the binomial Latin name *Hyperhydra egedei* (i.e., Egede's super-otter), in honor of its first sighting in Greenland. He includes a representative illustration for the typology (Figure 2.3), and notes that, as no sightings have occurred since 1848, the creature "may well now be extinct" (Heuvelmans 1968:548).

That the Egede UMO was a super-otter and not a run-of-the-mill cetacean is, according to Heuvelmans, demonstrated by examining Bing's depiction of its exhalation:

Robert L. France

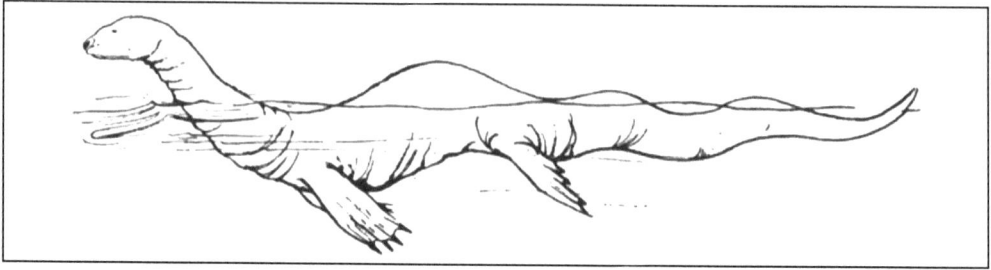

Figure 2.3. Heuvleman's (1968:547) suggestion for the identity of the Egede UMO, which he christens *Hyperhydra egedei,* or Egede's "super-otter," and which he believes may actually be a type of prehistoric whale.

> The animal's breath has only once been seen, by Hans Egede, the only sighting in a really cold area. The breath was emitted not from a vent in the top of the head, but from two nostrils at the top of the nose and was no doubt visible because of condensation as it came in contact with the very cold air. (The breath of cetaceans which 'blow' is visible in all latitudes; this is probably both a result of sudden decompression and because it contains droplets of oil.) (Heuvelmans 1968:547).

Somewhat confusingly, Heuvelmans offers the hypothesis that the Egede super-otter may have actually been a type of zeuglodon (Figure 2.4; see Chapter 5), a genus of primitive whale that has been extinct for 30 million years, and which displays otter-*like* traits, rather than being a giant otter, per se. To support this assertion, he presents an illustration of a zeuglodon reconstructed from fossils, that does indeed resemble Egede's creature, at least certainly more so in this regard than is the case for either Lee's giant squid or Oudemans' long-necked paleoseal. His own illustration of the Egede encounter (Figure 2.2) is a sneaky bit of slight-of-hand. Both the described breath exhalation and mottled skin are shown, but he deemphasizes the small dorsal protuberance through either an airbrush removal or by hiding it beneath the surface of the waves.

Modern Biology

As explored in France (2019a), biologists today have been able to parsimoniously explain many historic sightings of sea monsters/serpents in terms of known animals that went un-

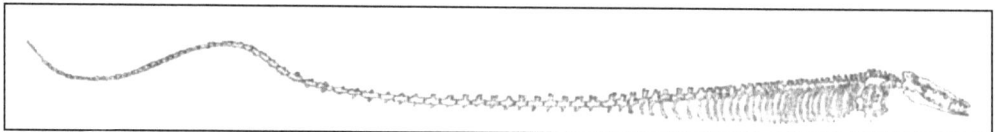

Figure 2.4. Fossil remains of a zeuglodon, a prehistoric whale which is believed by some, despite an absence of confirmatory evidence in the fossil record, to have escaped extinction to become the favored candidate proposed by many modern cryptozoologists, including Heuvelmans, for the Egede UMO (Oudemans 2007[1892]:327).

recognized at the time (e.g., Brongersma 1968; France 2016b, 2016c; Galbreath 2015). Sometimes this is accomplished through an objective procedure of comparative candidate screening (Paxton and Holland 2005; France 2017; Woodley et al. 2011). However, simply having modern biological training by no means *ipso facto* excludes the possibility that established scientific protocol is sometimes ignored and embarrassing blunders made, as for example, coining scientific names for the "Nessie" or an equally fictitious monster off the coast of British Columbia (LeBlond and Bousfield 1995; Scott and Rines 1975); this despite the absence of type-specimens as required by museums (Shermer 2003), and thereby being a process which has led to justifiable criticism (Naish 2012; Williams 2015). In contrast, other legitimate studies about marine mystery fauna often pertain to empirical investigations of either the likelihood of discovering news species of cetaceans or other large animals (Paxton 1998; Solow and Smith 2005; Woodley et al. 2008), or the gauging of the reliability of eyewitness accounts (Paxton 2009; Paxton and Shine 2016). And, as can be seen from the aforementioned citations, the leading modern biologist undertaking scientifically defensible research on UMOs is Charles Paxton.

Charles Paxton is an ecologist with an active research program focusing on, among other things, marine mammals. He is also the leading modern biologist engaged in publishing peer-reviewed articles in respectable academic journals on the question of sea serpents. In their 2005 paper, Paxton and his coauthors (E. Knatterud and S.L. Hedley) follow Thomas (1996) in also presenting an overview of the inaccuracies in the 1745 English translation of Hans Egede's encounter with the mysterious sea creature, before they move on to propose a novel explanation. The authors draw attention to the "strange protuberance" visible halfway down the body, which they suggest could be a dorsal fin, depending on which side of the body is being shown. Citing Egede, who stated that the creature was three to four times the size of the ship, and a Scandinavian publication estimating the size of such eighteenth-century vessels, Paxton et al. calculate that such the creature would have been between 64 and 98 m long. They conclude this to be "an implausible estimate of length" (Paxton et al. 2005:3), and one not in agreement with Bing's illustration which shows the UMO to have been only twice the length of the ship. They neglect to calculate that this in turn would still mean that the creature was still of an enormous size, between 42 and 48 m in length. This is an important point that I will return to later in my own reinterpretation of the UMO.

Paxton et al. mention that although Egede referred to the UMO as a "creature" and not a whale, "things that blow like whales are, all other things being equal, most likely to be cetaceans" (2005:3). The "carapace" or "shell-work" reads to them like barnacle encrustation, the single pair of fins observed suggests a whale rather than a large fish, and the backward plunge resembles standard cetacean breaching behavior. To counter the overt confirmation bias of cryptozoologists who promote their own pet theories without consideration of possible alternatives, Paxton et al. compared one behavioral and six morphological traits of the Egede UMO to those known for six species of large cetacean candidates. They conclude that the UMO "was likely to have been" (Paxton et al. 2005:1) either a humpback (*Megaptera novaeangliae*) or a North Atlantic right whale (*Eubalaena glacialis*), or perhaps one of the last Atlantic grey whales (*Eschrichtius robustus*) seen before the subspecies became extirpated from the region. The latter would explain, they believe, the eyewitnesses' unfamiliarity with the rare creature as well as the protuberance pictured, which might have been the dorsal fin of this particular species.

Finally, Paxton et al. address the perplexing issue of the serpent-like tail which is obviously unlike that for any known cetacean. They cite a paper from 1950 in which a grey whale was found to be missing its tail flukes, but correctly note that in such a case it would have been impossible for the Egede animal to have projected itself out of the water without a tail to provide the required force and thrust. Their alternative hypothesis is a most intriguing one, which has gone on to garner considerable press reportage.

Due to their large size and absence of clasping limbs, many male baleen whales possess long, serpent-like penises that are actually prehensile and thus can be twisted and turned about in order to facilitate copulation. Normally tucked up within the body to maintain a fusiform shape, when aroused, the penis can be quite visible. Several photographs of serpentine penises elevated above the water are presented in support. Paxton et al. state that such a penis "could be taken by a naïve witness for a tail" (2005:8). And there is no denying that silhouettes of such penises can look almost identical to Bing's close-up drawing of the Egede UMO's "tail." The inconvenient truth concerning the disparity in size between such penises, which are about 2 m in length, and the Egede creature's tail being the size of an entire ship's length from the body, is explained away as "the [possible] presence of more than one male whale" (Paxton et al. 2005:8). The authors suggest that such extended penises may explain at least one other famous sea serpent sighting (discussed in the next chapter), but offer caution that the Bing illustration may not be entirely accurate due the shown presence of teeth which would obviously preclude their baleen whale interpretation. However, this is not necessarily disqualifying, as toothed whales such as orcas (*Orincus orca*) have similar prehensile penises (albeit admittedly of smaller sizes). Paxton et al. also note that Egede senior, in his natural history writing about whales, does mention the large "*membrum virile*" of a whale, which implies that he and his son, who grew up in Greenland, would have been familiar with. But perhaps, the authors suggest, "the Egedes may not have realized that it could be seen at sea" (Paxton et al. 2005:8). Finally, Paxton et al. conclude that although all evidence points to the Egede UMO being a whale, we may never know for certain what particular type of whale was seen, the only certainty being that "it was a most unusual sight both at the time and now" (2005:8).

Modern Wordsmithery

As discussed in France (2019a), the appetite of the Victorian public concerning prehistoric sea dragons and contemporary sea monsters/serpents was insatiable. Every new set of bones dug up or mysterious creature spotted fueled considerable press reportage (Burns 2014). "Dinomania" (*sensu* Torrens 1991) ensued, leading to museum reconstructions, artistic interpretations, and exhibitions (Rea 2001; Rudwick 1992; Rupke 1994; see Appendix 1.1 and 1.5), and to the later development of "prehistoric- or crypto-fiction" wherein modern-day protagonists were portrayed battling gigantic marine animals that had escaped extinction by being sequestered either in remote locations around (as imagined by Arthur Conan Doyle and Edgar Rice Burroughs), or within (as imagined by Jules Verne and Edgar Rice Burroughs), the globe (Angenot and Khouri 1981; Mullis 2019).

Today, sea monsters/serpents continue to be a popular fixture of contemporary cryptozoology (Loxton and Prothero 2015; Rossi 2016) and culture (Hackett and Harrington 2018; Marven and James 2004; see Chapter 5 and Appendix 1.8). Indeed, one can choose from more than 4000 books on the subject that are available for purchase through the world's largest online retailer. One particularly engaging and memorable such offering in this regard is that by Bob Eggleton and Nigel Suckling.

The Book of Sea Monsters is a 1998 collaboration between artist Bob Eggleton, known for illustrating science fiction and fantasy books, and Nigel Suckling, a writer of biographies of fantasy artists. Eggleton's beautiful renderings are very fanciful, whereas Suckling's text is a barebones recounting of some of the more famous sightings. The Introduction begins:

> There are literally hundreds of plausible monster sightings on record; but some stand out in particular, because of the glaring honesty of the witnesses, the strangeness of the tale or simply the impact of that tale on the public imagination. An instance that scored on all three counts was the famous Egede sighting (Eggleton and Suckling 1998:7).

A more or less accurate version of the encounter is presented, followed by a discussion about the accurate version of the illustration cribbed from the original map. It is this which Suckling states has "caused the enduring fuss" (Eggleton and Suckling 1998:8). He notes that one of the interpretations advanced to explain the illustration is that it might depict a large seal "wrestling" with a squid, but does not provide a reference. He notes—again without background support—that the "most popular theory" is that it shows, *à la* Lee, one tentacle and the "tail" (whatever structure that might be) of a large squid. Suckling is correct in noting that such an explanation does not account for the large flippers, and that there is no squid of such a size that is capable of showing fully 9 m of its body or mantle above the water. This leads him to conclude that "so this 'rationalization' in itself ends up arguing for a monster that is beyond the pale of science" (Eggleton and Suckling 1998:8). Near the end of the book, in a chapter entitled "Red Herrings," based on the idea that "many sightings of sea monsters can still be easily explained in terms of the known" (Eggleton and Suckling 1998:70), Eggleton offers, as a modern equivalent of Hamilton's illustration, his own imaginative interpretation of the 1734 sighting (Figure 2.5).

Figure 2.5. The artistic imagination of Bob Eggleton on full display with respect to the Egede UMO in his and Nigel Suckling's coffee-table collaboration, *The Book of Sea Monsters* (1998:71).

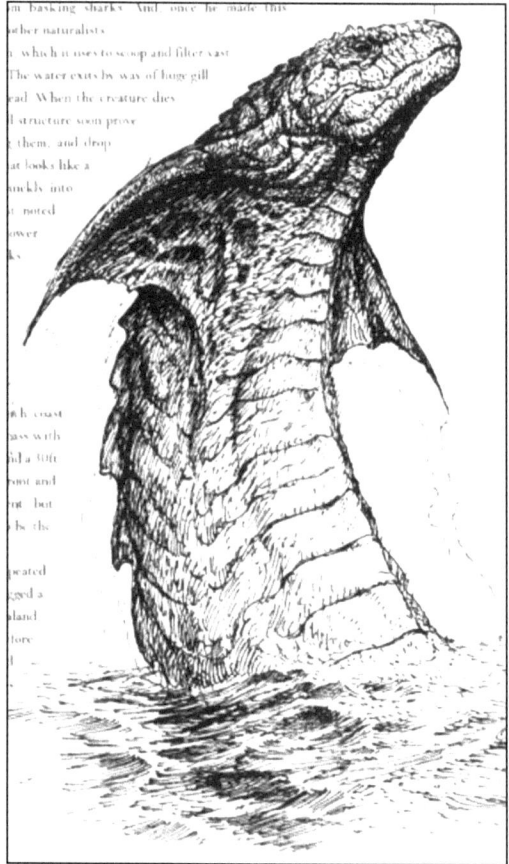

Supplementary Expositions

Early Natural Historians

The German Moravian missionary, David Cranz, was a contemporary of Poul Egede. The English publication of his *The History of Greenland* in 1767 (with his name altered to "Crantz") was highly influential (e.g., Samuel Johnson referred to it, and Joseph Banks and Captain James Cook both carried it on their respective voyages). As Jensz (2012:460) states: "Both Cranz's and Egede's books found large popular audiences in England and also Germany," introducing the words "*kayak*" and "*umiak*" to Europeans. Regarded as a classic of Protestant missionary literature, the book is a hybrid of travelogue, non-devotional religious text, and ethnographical treatise. Cranz was a highly capable naturalist, dedicating entire chapters to meteorology, geography, oceanography, botany, ornithology, in addition to marine biology and ichthyology. Interestingly, his descriptions of mysterious marine fauna in Greenland are also thought to have inspired Coleridge's choice of the words "slimy things" in the *Rime of the Ancient Mariner* (Jensz 2012; Ower 2001). Accurate descriptions of cetaceans and pinnipeds are provided (though no mention is made of the, by then, possibly extirpated Atlantic grey

Robert L. France

36

whale). Cranz recounts Egede's descriptions of the kraken, merfolk, and the "most dreadful monster." He is circumspect with regard to advancing his own theories, instead defaulting to Pontopiddian, excerpts from whom are quoted at length in accompanying endnotes. In this regard, Cranz supports the identification of the Egede creature as a sea serpent, albeit this is expressed with a fair degree of skepticism that is well in advance of its time, as reflected by his belief that the imagination of native Greenlanders with respect to black bears had "exaggerated them into monsters" (Cranz in Parish 2020:6; or perhaps this may be a consequence of cultural differences; i.e., there is a marked paucity of Germanic compared to Nordic and British accounts of sea monsters). Near the beginning of a copious endnote, Cranz refers readers "who wish to feed their love of the marvellous with a few of the fanciful creations of our forefathers, are desirous of a subject for the exercise of innocent ridicule" (Crantz 1767:323) to Pontoppidan. And despite recognizing the broad appeal of the Sea Serpent on cultural imagination, noting that it is "a creature proportionate in size to the extent of its domain" (Crantz 1767:329), Cranz continues by offering tacit support for the theories of the bishop of Bergen:

> Many of our readers are probably already tired of what may appear to them idle fictions, and we should not trouble them with anymore, if not the evidence of the writer above quoted [i.e., Pontoppidan], appear worthy of being more generally known and examined. Besides, he enters upon his discussion with such spirit and interest, manages his inquires with such precision, we had almost said philosophical correctness, that he really induces some degree of participation in his zeal for maintaining the reality of an existence, which, in itself, it but barely probable (Crantz 1767:329).

Back in the main text, Cranz explicitly identifies Egede's "enormous Sea-serpent" from Greenland with Pontoppidan's Norwegian sea-orm. And in an illuminating section, prescient to the novel illation posited in the next chapter, Cranz sums up the species:

> Their length is estimated at a hundred fathoms [i.e., 180 m, a biologically impossible length almost twice that of the largest whales], and their thickness at about two yards. Their convolutions [i.e., humps], which are from twenty to a hundred, look like large hogsheads [barrels] floating on the surface of the water. The northern poet, Peter Das [cited in Pontoppidan], has a simile, in which he compares them to a hundred heaps of manure, laid out in order on a field, and gives them the epithets of Behemoth and Leviathan (Crantz 1767:121).

It is hard not to be impressed by Otto Fabricius, the eighteenth-century Danish missionary, cleric, philologist, naturalist, and ethnographer. Schooled by Poul Egede, Fabricius would later be elected to membership of the Royal Danish Academy of Sciences and Letters. That he left the European colony to live on an intimate basis with local Inuit for half a decade is admirable. It is during this period of becoming a seal hunter that Fabricius assembled, from personal experience of direct observation, the detailed zoological and ethnographic information that would be published in his masterpiece *Fauna Grøenlandica* (1780), and a decade later, his equally impressive *The Seals of Greenland* (Kapel 2005). Kapel (2005:19) refers to these monumental efforts as "standing in a class of their own … tower[ing] high above contemporary faunistic works," leading him to conclude that "it is possible to say that scientific knowledge of the fauna of Greenland begins with Otto Fabricius." Fabricius also wrote a suite of scientific papers on a wide variety of subjects, such as a treatise on drift ice.

Written in Latin, *Fauna Grøenlandica* (1780) describes 473 animal species from Greenland, 130 of which—mainly small invertebrates—being new to science. The strength of Fabricicus' books lie in the fact that rather than being simple inventories, they contain considerable biological information. So not only does he report on the occurrence of 15 species of whales, he gives meticulous details of anatomy and behavior for some of these that are so accurate they would not to be confirmed more than until a century later. For example, he provides some of the very first descriptions of humpback breaching, something that led him to being dismissed by later naturalists as being a fabulist, since no logical person could ever countenance that any 17 m long creature could be capable of being be able to jump completely out of the water (Kapel 2005). Fabricius even suggests that breaching might be a cetacean strategy to dislodge ectoparasites, something widely regarded as a valid hypothesis today. Moreover, he provides what might be the first mention of cetacean mortality resulting from the infection of harpoon wounds of individuals whom had managed to avoid capture. And important from the perspective of the present investigation is that Fabricius, through his close working relationship with native Greenlanders—as, for example, his descriptions of their use of inflated seal-skins as floats during harpoon hunting (a topic revisited in Chapter 3)—was sensitive to local knowledge of rare species. In consequence, there was no one better positioned to be able to knowledgably comment upon the mysterious creature of the Egedes. Unfortunately, and quite remarkably, he offers no suggestions as to the nature of the beast; nor does he even mention it. Fabricius' silence in this regard is notable, suggesting the possibility of deliberate omission as a way not to embarrass his missionary predecessors.

The Naturalist's Library was a series of 41 books published between 1833 and 1866 by Sir William Jardine, the Seventh Baronet of Applegarth. The collection served as the touchstone of information about the natural world and became *de rigueur* for all gentlemen natural historians (Sheets-Pyenson 1981). Robert Hamilton, M.D., penned two volumes: one on fishes, and a second, in 1839, entitled *The Natural History of the Amphibious Carnivora: Including the Walrus and Seals, and the Herbivorous Cetacea, Mermaids, etc.* In the chapter "The Great Sea-Serpent," Hamilton reviews the Gloucester sightings (France 2019a, 2019b), Stronsa Beast carcass (see Chapter 5), Pontoppidan's creatures, and speculations about the legendary kraken. Short shrift, however, is made of the Egede encounter. After presenting the English-translated "paws" version, Hamilton describes it as an "unquestionably exaggerated statement of the honest missionary, Hans Egede, concerning what he tells us he himself witnessed off the coast of Greenland in the year 1734" (1839:429). The choice of wording used—"he tells us"—is disparaging of the eyewitness's capabilities. This rebuke by Hamilton is somewhat ironic given that the fanciful illustration of the encounter by James Stewart that accompanies the text (Figure 1.12), has done more to single-handedly create the enduring myth of sea serpents than perhaps any such other.

Intriguingly, and in keeping with the hidden nature of cryptids, when one consults the online Google-digitized version of Hamilton's (1839) volume made from a copy in Oxford University's Bodleian Library, and flips through the dozens of gorgeous paintings of wildlife by Stewart, Plate XXVIII, that of the Egede UMO, is absent, having been ripped from the volume at some time in the past. If of an ethnozoological bent, this bespeaks of the fascination that people have with sea monsters (see the Appendix); if, however, of a cryotozoological persuasion, this is the material from which conspiracy theories about truth suppression are born.

Philip Henry Gosse was already a well-travelled natural historian and popular writer, as well as the inventor (of both the word and the physical entity) of the aquarium, when he penned, in 1860, *The Romance of Natural History*. Devoutly religious, today he is best remembered for his writings trying to reconcile concepts of geologic deep time and Biblical chronology. Gosse is also important for being one of the first amateur natural historians of reputable experience about marine biodiversity to speculate on sea serpents. Notably, however, despite providing an historical overview of evidence for the existence of sea serpents, he fails to even mention the Egede encounter, possibly believing it to be unreliable. However, the glaring absence of referral to the 1734 UMO in Gosse's work is perhaps a consequence of it being too difficult to fit those observations into his own pet theory that,

> there exists some oceanic animal of immense proportions, which has not yet been received into the category of scientific zoology; and my strong opinion, that it possess close affinities with the fossil Enaliosauria [i.e., an obsolete term for the prehistoric reptiles plesiosaurs and ichthyosaurs—see Appendix 1.1, 1.3, 1.5, 1.6] of the lias [i.e., the Liassic or Jurassic Period] (1860:344).

I've included this absence of mention in one of the seminal Victorian books on natural history because such selective screening of data would go on to become a standard *modus operandi* of sea monster cryptozoology (France 2019a).

Robert Brown was a noted Victorian scientist, university lecturer, and explorer, the latter taking him as far afield from Scotland as Spitzbergen, Venezuela, the Pacific Islands, Siberia,

Vancouver Island, North Africa, and Greenland. Seemingly being one unable to relax, in addition to his globetrotting, he authored six books (several in multiple volumes) as well as dozens of academic papers on geology, botany, and zoology. Despite writing with some sensitivity about the dramatic alterations in the lives of Indigenous peoples of the Pacific Northwest during the nineteenth century, Brown, as noted above, was critical of, in his mind, Fabricius' misguided trust in the acumen of the Greenland Inuit in regard to their distinguishing among marine fauna. His 1868 paper on the biology of pinnipeds in Greenland and Spitzbergen became the standard piece of scholarship on the subject for many decades. It is in another paper from the same year—"On the Mammalian Fauna of Greenland"—where Brown demonstrates that, steeped as he is in the scientific realism of the mid-Victorian Age, in his mind there is simply no place for fuzzy logic and lore from the past. He is dismissive of the earlier natural histories of Greenland by both Egede and Cranz as being unsatisfactory. As such, Brown will quite simply not even countenance the existence of the unexplained. He dismisses the idea that the great auk (*Alca impennis*) could have escaped extinction by hiding out in a remote location on the island; nor, in a section entitled "On some of the doubtful or mythical animals of Greenland," does he give credence to the Egede UMO: "Still less I will stop to inquire about that 'sea monster' which good Paul Egede saw, and Pastor Bing sketched 'off our colony in 64° north latitude'" (Brown 1868:362). He concludes the paper with the bold statement that,

> I have said enough to show that, though, there is yet much to be done to the legitimate zoology of Greenland proper, there is still more to be done in what may be called the illegitimate zoology—the history of zoological myths and errors (Brown 1868:362).

By this, Brown presumably means addressing taxonomic mysteries through what is today regarded as the discipline of folk-biology or ethnozoology, an enterprise whose veridicality of which he seems to hold a pejorative opinion.

John George Woods was a well-known (enough to be quoted in fiction by both Mark Twain and Arthur Conan Doyle) Victorian parson and a popularizer of natural history. This comes from him being the prolific author of many books, one of which sold thousands of copies in a single week (outselling Dickens' latest novel no less). In his 1884 article for the *Atlantic Monthly* magazine, "The Trail of the Sea Serpent" (Woods 1884), he reviews many famous sea serpent sightings and refers to the "remarkable narrative" of Hans Egede. Wood makes repeated reference to the "distinctly delphinian" or dolphin-like traits of the illustrated UMO, leaving little doubt that he considered the Egede creature to have been some hitherto unseen variation of that type of mammal.

John Gibson's *Monsters of the Sea, Legendary and Authentic* was published in 1887 during the period of Victorian fascination with the possibility that the seas might contain primordial creatures. Five chapters focus on the subject, the first beginning with the definitive statement "That sea-serpents thus exist in abundance is beyond all doubt" (Gibson 1887:42). This is actually one of the very few times where Gibson expresses a personal opinion. His text is a factual retelling in his own words of various first-hand encounters with mystery animals, including Egede's creature. This is accompanied by a facsimile of Pontoppidan's illustration. No specific comments are offered apart from reiterating Lee's recent theory of the animal having been a giant squid. Such an absence of offering ill-founded conjecture as to the UMO's identity is a rare caution in the writings of amateur Victorian natural historians, and I've included this example for that reason.

Victorian Natural Scientists

Edward Newman was an entomologist, botanist, and editor of the important Victorian journal of natural science, *The Zoologist*. Recognized today for being one of the first to propose that certain dinosaurs might have been warm-blooded, Newman was also fascinated by the possibility of sea serpents, frequently using his journal to comment upon the subject. One such contribution from 1849 was "The Great Sea Serpent: An Essay, Showing its History, Authentic, Fictitious, and Hypothetical," in which he praises the accuracy and fidelity of Egede's abilities as a natural historian, before offering the following synopsis:

> It seems to us indisputable, that Mr. Egede, from personal observation, and with rigid intensity of purpose, describes and figures an animal decidedly and wisely different from any living creature hitherto admitted into our systematic classifications. That it was a sea-serpent, or a serpent of any kind, certainly does not appear, neither does the writer make any such assertion (Newman 1849:1605).

Basing his commentary on the inaccurate translation, Newman states that the use of the term "sea monster" has allowed "this very name [to have] been tortured into a proof of the falsehood of Mr. Egede's statement" by skeptics (1849:1606). For Newman, the mystery of the Egede creature, and a good many other such sightings, can be explained as them being surviving plesiosaurs, similar to the fossils that were being discovered and displayed at the time (see Appendix 1.1 and 1.5). This is a classic example of the faddism that often characterizes sea-serpentology (see Paxton and Naish 2019).

Armand Landrin was a French anthropologist and curator at the Museum of Ethnography in Paris. Having training and experience in both geology and biology, he felt called upon, as did many such educated scholars at the time, to offer his own comments on the exciting and perplexing accounts of sea monsters that had become prevalent. In his 1875 *The Monsters of the Deep, and Curiosities of Ocean Life: A Book of Anecdotes, Traditions, and Legends,* Pontoppidan's representation of the Egede creature is presented, followed by a brief discussion in which Landrin suggests the animal to have been some type of marine reptile, presumably an hitherto unknown species of giant snake.

Son of the leading ornithologist of the day, and thus raised with a zoological education, Charles Gould, in his 1886 *Mythical Monsters,* reviews the cases made for the existence of such imaginary beasts as unicorns, dragons, and the phoenix, before moving on to conclude with a long chapter on the nature of sea serpents. Using the inaccurate "sea-monster … blew water … broad paws … scales" translation and accompanying illustration, Gould concludes that such traits preclude the creature from being a serpent. Noting that the Egede creature was similar to others sighted elsewhere, he suggests the animal to have been a prehistoric saurian from the age of great sea reptiles in the Liassic Period (today referred to as the early Jurassic; see Appendix 1.1, 1.2, 1.3).

William Hoyle was a British biologist who assumed the directorship of several major museums during his distinguished career. His specialty was marine animals, particularly cephalopods, and he is best known for describing species from the famous 1872–1876 *Challenger* scientific expedition. Part of the expedition's mandate was to test the theories of the notable natural scientists Louis Agassiz and Thomas Huxley by sampling the deep ocean for "living fossils" or "missing links" which possibly could have survived in those locations due to the theory that stable conditions slowed down evolution; i.e., in contrast to the extinctions which have occurred in the more dynamic environment of shallower waters. This expertise led to Hoyle being invited to contribute the entry on sea serpents in the 1902 edition of the *Encyclopedia Britannica.* Following introductory mention of Magnus and Pontoppidan and the Nordic origin of sea serpents, Hoyle lists nine explanations to account for "the causes which presumably gave rise to the phenomena described" (1902). Most of these are mundane, such as the common behavior of several basking sharks swimming in a row, snout-to-tail. And, not surprisingly given his own area of zoological expertise, Hoyle reiterates Lee's hypothesis that the Egede monster was a misidentified giant cephalopod. To emphasize this, he includes Lee's illustration of a squid raising itself out of the water for reinforcement. The only other graphic shown is that of the 1848 *Daedalus* sighting, which Lee had originally, and now Hoyle again, suggests may have also been due to an unrecognized giant squid. Such prominence given to the Egede creature in the world's leading English-language dictionary of accumulated knowledge indicates that even more than a century-and-a-half after its sighting, the events were still being regarded as one of the most significant of all such encounters.

Hoyle is also notable for remaining open-minded about what was or was not known about marine biodiversity at the time, concluding that,

> it would thus appear that, while, with very few exceptions, all the so-called 'sea serpents' can be explained by reference to some well-known animal or other natural object, there is still a residuum sufficient to prevent modern

zoologists from denying the possibility that some such creature may after all exist (1902).

Cryptozoologists

Commander Rupert Gould had retired from a distinguished naval career when, in 1930, he wrote *The Case for the Sea-Serpent*. Gould's book is often regarded as "a model of scientific rigour" by cryptozoologists (Heuvelmans 1968:440). Part of this is due to his approach in focusing on just a handful of UMO sightings from the entire corpus, these being specifically selected for their reliability compared to the bulk of accounts.

Gould begins his work by examining the Egede sighting, but does so with a clearly stated skepticism, something notably absent from much subsequent cryptozoolgical conjecture by others. Despite not "quite reaching the standard of evidence aimed at in the succeeding chapters," one has to seriously consider the encounter, he believes, due to its foundational "importance in the history of the subject [of sea serpents]" (Gould 1930:11). Gould, too, follows Oudemans in pointing out the various distortions in the text and illustrations concerning the sighting. He notes that the evidence that Reverend Bing was present at the sighting is presumptive but not quite conclusive, thereby calling into question how soon after the encounter the illustration was made. This is important, Gould suggests, as interpretations based on the illustration "can scarcely be regarded as carrying the same weight as Egede's description" (1930:15). Unaware that it is possible that Egede senior may not have himself been an eyewitness to the encounter (see Thomas 1996 below), Gould provides a character defense of the Reverend's truthfulness as a reason for why we should trust what he wrote.

At first glance, it would seem that the leaping up and then backward plunge of the UMO implies it to have been a breaching whale. However, Gould notes that elsewhere in Egede's *Description of Greenland* (1745), various species of whales are described "very clearly and accurately" (1930:16). This is significant in Gould's mind given Egede's purposeful use of the words a "very terrible sea-monster" (actually, as has been seen, "sea-creature" is the more accurate translation), and the expression that "it blew like a whale." Obviously, Gould argues, a man so well acquainted with whales as Egede was would not use such language (i.e., "*like* a whale") if he believed the UMO to have truly *been* a whale. This is an important point to keep in mind when I advance my own alternative hypothesis for the animal behind the UMO in Chapter 4. Gould then goes on to liken the resemblance of the Egede creature to that of an UMO observed in 1891 off the coast of New Zealand. He therefore concludes that Egede's monster "was not, it would seem, a whale; and it was certainly too large for a shark or any other known creature" (Gould 1930:17). He dismisses Lee's hypothesis, noting the latter's cleverness in making his drawing of a giant squid to resemble Bing's original illustration, but stating that the idea "will not stand examination" (Gould 1930:20). He reminds the reader that Egede's description has the creature blowing like a whale from the head, "not from a submerged tube some thirty feet distant." And like Oudemans, he too comments on the mechanical impossibility of the mantle and tentacles being simultaneously raised above the water to the heights observed. All this leads Gould to sum up inconclusively, albeit cautiously, that "Egede saw a large creature of serpentine form and unknown type" (1930:21). This particular example of Gould is significant in showing that it is possible for non-professional biologists to remain both open-minded and critically skeptical with regard to cryptozoological assumptions and commentary made about mystery animals.

Fortean Studies was a magazine that published articles, some critical, about fringe science and anomalous phenomena. In his 1996 article therein, Lars Thomas, originally trained as a zoologist but now operating as a cryptozoologist through authoring books on the natural history of trolls and the like, takes issue with Heuvelmans' super-otter categorization of the Greenland UMO, through noting that it was based on the inaccurate English edition of Hans Egede's book. To counter this, he presents the original Danish and correctly-translated versions from both son and father, along with a cropping of the illustration of the creature from Bing's map. Thomas makes the following points: 1) the drawing shows a "small, but very distinct dorsal fin" (1996:236); 2) had the creature a second, posterior pair of flippers, it would have been noted when it hurled itself backward; and 3) the skin cannot be likened to fur of any kind but rather "sounds more like the skin of a grey whale, covered with barnacles and other forms of growth" (1996:236). Thomas considers Egede's creature to have been an "entirely different animal to the Super-otters Heuvelmans records as having been seen along the coast of Norway" (1996:236). And in a cryptozoological flourish, he concludes that perhaps the scientific name for such super-otters should be changed to *Hyperhydra norvegica*. As to what Thomas thinks the Egede creature might have actually been, the only hint given is that he estimates the size of the creature to be from 30 to 35 m in length, and remarks that it "looks very much like the zeuglodont … [an] ancient form of whale so far known only from fossils" (1996:236).

Karl Shuker is a zoologist who has published a number of books about the documented discovery of legitimately described new species. He is also the founder and editor-in-chief of the *Journal of Cryptozoology,* and was a frequent contributor to *Fortean Times*, a magazine that dealt with anomalous phenomena. In addition to his many published books on all kinds of cryptids, his 1996 *In Search of Prehistoric Survivors: Do Giant 'Extinct' Creatures Still Exist?* continues to be well-regarded in the cryptozoological community. An accurate version of the Egede UMO is presented therein, but as he is unaware that the textual translation being worked from is faulty, he spends a paragraph arguing that one should not take the front limbs to literally be "paws" when they are really fins. Influenced by the paleontological discovery the previous year of an *Ambulocetus* or "walking whale" (i.e., whales are known to have evolved from land animals), he suggests that a survivor of this most primitive cetacean would explain the paw-like flippers of Egede's UMO. However, he does not explain how that a sea lion-sized prehistoric creature could account for the large size of the animal seen in Greenland, nor how it could have escaped being present in the fossil record for millions of years. It is enough that the ancient whale has four short limbs and large feet (Figure 2.6) and therefore looks more like a crocodile, hippopotamus, or a pointy-nosed giant otter than a whale, per se. In terms of the latter, Shuker suggests that the discovery of *Ambulocetus* provides support for Heuvelmans' hypothetical super-otter. He does not completely rule out the identity possibility of a zeuglodon, however, stating emphatically that "I simply cannot see why at least one lineage of zeuglodonts could not have persisted into modern time"

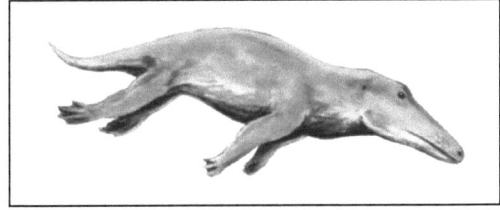

Figure 2.6. An *Ambulocetus,* an identity candidate advanced by Karl Shuker in *In Search of Prehistoric Survivors: Do Giant 'Extinct' Creatures Still Exist?* (1996) for the Egede UMO. If this most ancient type of prehistoric cetacean somehow escaped extinction (and detection due to the imagined failure of modern science), it gives credence, so argues Shuker, to Heuvelmans' super-otter hypothesis.

(Shuker 1996:113). He goes on to argue that the absence of discovery of such living creatures today is not because they do not exist but is rather due to the failure of science to devise the correct means of finding them. There is not enough space in the present monograph to fully address the absurd tautology of such statements of cryptozoological logic, but see a detailed, well-referenced argument about the egregious weaknesses in the pseudoscience in France (2019a). This example, like the previous one, is informative in demonstrating that a pedigree of a modern education in biology does not preclude the generation of conclusions that are unscientific or "abominable" (*sensu* Loxton and Prothero 2015).

Despite the complete absence of any physical remains of sea serpents, and therefore an inability, as per appropriate scientific protocol (Loxton and Prothero 2015; Shermer 2003), to formally identify and legitimately register the discovery and existence of such creatures, cryptozoologists seem preoccupied with developing elaborate classification schemes for the imaginary creatures. Engaging in such endeavors is what I have referred to as producing a "nomenclature of nonsense" (France 2019a). One of the most comprehensive of such house-of-cards frameworks is presented in Loren Coleman and Patrick Huyghe's popular 2001 book, *The Field Guide to Lake Monsters, Sea Serpents, and Other Mystery Denizens of the Deep.*

Coleman and Huyghe believe that Heuvelmans' super-otter *Hyperhydra egedei* category is best ignored as it is predicated on the inaccurate translation as noted by Thomas (1996). Careful reading of the original account generates, the authors believe, a "description [that] sounds far more like an ancient whale than anything resembling Heuvelmans's Super-otter" (Coleman and Huyghe 2001:34). And so they proceed to lump Egede's UMO together with the famous cases of the sixteenth-century Scandinavian sea-orm and the nineteenth-century Gloucester creature, into the category of the many-humped "Classic Sea Serpent," of which they offer a representative illustration (Figure 2.7). However, perhaps somewhat unsure about it all, in the end they notably back away from specifically flagging the Greenland encounter

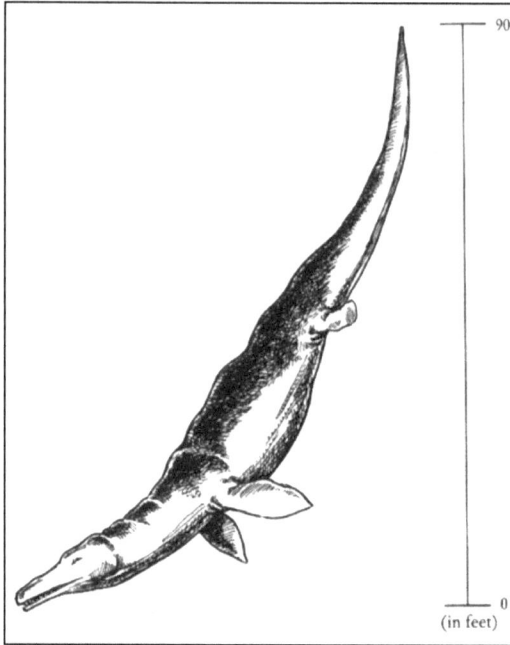

Figure 2.7. Illustration by Harry Trumbore of Loren Colman and Patrick Huyghe's supposition of the archetypal or classic sea serpent which they consider, in *The Field Guide to Lake Monsters, Sea Serpents, and Other Mystery Denizens of the Deep* (2001:49), to explain many sea serpent sightings, including the Egede UMO, and which they posit might be zeuglodons having escaped extinction.

as being one of the six clarion examples that best illustrate this particular category of their classification scheme.

George Eberhart is a library researcher and published "ufologist," whose 2002 book *Mysterious Creatures: A Guide to Cryptozoology* provides a compendium of mystery animals from around the world. With respect to the Egede UMO, its "long tail that ends in a point" leads him to suggest that it might have been an archaic whale "more primitive than the basilosaurids, with some vestigial limbs" (Eberhart 2002:530). Not stopping there, he further speculates that these cryptids were recently distributed across the Arctic Ocean, from Greenland to Norway, and occasionally within the Baltic Sea. He also believes the species was last seen in Norway in 1847 before becoming extinct shortly thereafter.

Michael Woodley is a biologist who conducted doctoral research on community ecology and who presently studies the evolution of human cognitive behavior. He is also the author of several papers challenging the supposition that certain UMOs might be hitherto unknown cryptids. His 2008 *In the Wake of Bernard Heuvelmans* was published by the Centre for Fortean Zoology, a cryptozoology society that has produced other books on dragons, enormous cats, monstrous birds, British big-foot, and the ilk. Woodley, accepting the existence of aquatic cryptids, provides an overview of the history of sea serpent classification, followed by his own suggested refinements. With respect to Egede, the erroneous illustration of the water-spouting UMO is presented along with the inaccurate "monster and broad paws" translation of the encounter. His text mixes a review of Heuvelman's theory that the Egede UMO was a super-otter, with descriptions about the biology and ecology of real sea otters, before moving on to dismiss Paxton et al.'s theory:

Although it is an interesting hypothesis, and could certainly account for some sea serpent sightings, it is ambiguous as to how well it fits the facts in this case. The scallop patternation that Egede describes on the body of the creature that he saw, is consistent with that found on certain cetacean penises; however the size of the creature observed by Egede, and the presence of facial characteristics are clearly inconsistent with this hypothesis. For the Egede cryptid to be a whale penis, serious exaggeration has to be assumed on the part of Egede. As Huevelmans points out, Egede's writing style is too factual and objective. Additionally, his son, who was also a witness to the event, corroborated much of his report (Woodley 2008:100).

As an alternative, Woodley wonders whether Egede might have seen "a sea serpent of the super-otter variety" (2008:100). He cautiously states that,

> simply because the cetacean penis identity does not seem a likely candidate for his [Egede's] 'Most dreadful monster' does not mean that he was necessarily witness to something new to science (Woodley 2008:100).

And so, whereas the identity of the Egede creature is left as an open question, Woodley nevertheless retains Heuvelmans' super-otter category, though reassigning the animals to being true lutrinae (otters) rather than a primitive archaeocete (extinct whale) as had been originally proposed.

Lars Thomas expanded upon his 1996 thoughts about the Egede UMO in *Weird Waters: The Lake and Sea Monsters of Scandinavia and the Baltic States*, published in 2011 by the cryptozoological Centre for Fortean Zoology. The use of the word "Fortean" in all these cryptozoology publishing venues denotes the study of anomalous or paranormal behavior, such as that expounded by Charles Fort, whom holds an esteemed position in the developmental history of this field of fringe science in much the same way that Darwin and Olmsted do, respectively, to evolutionary biology and landscape architecture. Herein, he emphasizes that the numerous drawings and descriptions of various cetaceans in the Egedes' books indicates them to be very familiar with Greenland fauna, so much so that it is extremely unlikely for them to have failed to identify a whale. But some type of whale the UMO must have been, Thomas believes, given the UMO's size and the presence of a dorsal fin. Including the original Bing illustration cropped from the map, he states that it is the absence of the dorsal fin in subsequent renderings that sent Heuvelmans badly off track with regard to his super-otter theory. Neither does Thomas give credence to Paxton et al.'s penis theory, noting that the Egedes were "quite familiar with the male whale's rather awesome 'equipment'" (Thomas 2011:12). He is right to be puzzled by the fact that all known species of whales have nostrils located on the top of their heads and not at the tip of the snout as shown in Bing's illustration.

This leads him to reiterate his conclusion that "there is no possible candidate for the Egede beast" (Thomas 2011:12), expect perhaps the zeuglodon, a primitive whale he believes, just like the coelacanth (that favorite example that cryptozoologists frequently trot out to support their overt confirmation biases), that has escaped extinction. He does mention that until the creature is filmed or a carcass washes up somewhere, the mystery of this UMO "will never be solved" (Thomas 2011:12).

Most modern cryptozoologists, when they refer to the Egede UMO (and there are dozens examples of this online), do so by merely reiterating previous theories. Occasionally, however, some do advance novel interpretations that are worthy of serious consideration. One such in this regard is from Dale Drinnon, a frequent contributor to the blog run by the International Cryptozoological Community at the Centre for Fortean Zoology. In his April 2010 blog posting, he mentions discussing the matter of the Egede UMO with zoologist Charles Paxton at great length. Drinnon believes the Bing illustration to be of value in depicting the sighting, believing it to be "a plausible enough depiction of a gray whale." He demonstrates this by showing several photographs of breaching gray whales which undeniably do look remarkably like the front-end of the UMO in the Bing illustration. But the problem has always been the incongruous long tail, about which he states, Paxton et al. "guessed."

Drinnon (2010) goes on to state that,

> in the case of the tail I even noticed that the shape that was drawn was a fair depiction of ONE fluke of the tail (Only). Presumably the tail was viewed briefly and from such a position that it appeared to be folded over.

To illustrate this intriguing suggestion, he juxtaposes a photo of a breaching gray whale on one side with a cropped photo of another whale's tail (i.e., to show how it would look should one fluke be folded over) on the other. And he posts this fusion near a reprinting of Bing's original illustration to indicate the overall similarity. Drinnon concludes by mentioning the parsimony of his hypothesis (thereby demonstrating that not every cryptozoologist is guilty of ignoring the lessons of Occam's razor):

> So it boiled down to yet another fleeting sighting of an unrecognized but known animal. No need to make any extra special conditions or make any special arguments over it. Hundreds or even thousands of water-monster sighting [sic] the world over and throughout history fall into that description."

One respondent to the blog posting, however, remained unconvinced of the "known animal" advanced by both Paxton et al. and Drinnon, noting that gray whales simply do not have "the strange 'pectoral' fins" shown in the original illustration (2010).

Retired engineer and cryptozoologist Bruce Hynes' 2012 book *Here be Dragons: Strange Creatures of Newfoundland and Labrador* is, as the back cover declares, "dedicated to all those who claim to have seen something peculiar snaking beneath the swells or shifting through the shadows in our own backyard." The accurate illustration and the "most dreadful monster" mistranslated description of the Egede encounter are presented. This is followed by a curt and unexplained dismissal of the recent theory that the UMO might have been either a North Atlantic humpback, right, or one of the last grey, whales. Instead, rather than these so-called "unlikely candidates," Hynes advances, in typical cryptozoology fashion unsupported by any scientific evidence, that "this creature may have been a form of primitive cetacean, which has since become extinct" (2012:35).

Modern Biologist

Maurice Burton was a zoologist specializing in sponges, who also wrote popular science books, one of which was based on questioning the evidence about a certain large monster thought to exist in Loch Ness, about which he suggests several natural phenomena to be responsible for the sightings. He also contributed several articles to the debate about the nature of the famous "Soay beast" UMO, proposing it to have been a leatherback sea turtle (*Demochelys coriacea*; see France 2017). In his 1954 book, *Living Fossils,* he reviews the menagerie of strange but real animals that have escaped extinction to persist for millennia. The final chapter, "Monsters and Mystery Animals," includes comments about many such well-known animals that continue to fuel to cryptozoological beliefs. Herein, he skeptically dismisses the bulk of sea serpent sightings as being "a welter of optical illusions, practical jokes, hoaxes and imperfect observations" (Burton 1954:232), but does admit that some accounts should be treated seriously. He goes on to conclude that some hitherto unrecorded type of "giant eels were probably responsible for all the stories about sea-serpents" (Burton 1954:232), including Egede's UMO. This contention is based on the behavior and physiology of known species of eels, and the at-the-time recent discovery of an elongated larva of one such that had been collected from great depth in the ocean, which suggested it might have grown to an adult of monstrous proportions.

Wordsmiths

John Ashton was a nineteenth-century British author of 30 books, most being histories of the Regency Period. From his desk in the Reading Room of the British Museum, Ashton researched and wrote about topics as broad as Christmas traditions, city parks, other writers, and social mores. Besides his 1890 book *Curious Creatures in Zoology*, the closest he comes to the monstrous is in editing the *Travels of Sir John Manderville*, which he annotated and illustrated. In his Victorian bestiary of some real, but many fabulous animals, Ashton includes a chapter on "The Sea-Serpent," wherein he touches upon its representation in Assyrian sculpture before moving on to recent times. He states a belief (Ashton 1890:269) that:

a) such creatures "seem to be have been seen in more northern waters;" and b) many people no longer report their sightings for fear of being ridiculed given that skeptics can be assured to rush forward to advance their own pet theories about "a school of porpoises, or an enormous cuttle-fish, with its tentacles playing on the surface of the water" (i.e., the latter, a slight dis of Lee). A handful of famous illustrations are shown, including a cropping of the inaccurate version of the water-spouting Greenland UMO, about which his only comment occurs when he cites Magnus and Pontoppidan, noting that the latter's description of the sea serpent is "somewhat similar to Egede's" (Ashton 1890:271).

Willy Ley was a science writer who is best known today for being an early advocate of manned spaceflights. His 1948 book, *The Lungfish, the Dodo, & the Unicorn: An Excursion into Romantic Zoology,* is in the Victorian tradition of Philip Gosse's and Charles Gould's similarly themed works, and focuses on mythological, extinct, and relict animals. In the chapter entitled with the now standard catch phrase (adopted from Gosse) "The Great Unknown of the Seas," he introduces some of the better known sightings of sea monsters. Stating that it is important to go back to the origin of the phenomena, he reviews the Egede encounter, commenting upon the weather at the time of the sighting, while considering the Reverend to be the most reliable of witnesses. Because of Egede's familiarity with Greenland cetaceans, Ley concludes "his serpent cannot have been a whale, as he suggested" (1948:62). Nor, in his mind, had the UMO anything in common with the mythological Nordic sea-orm. Instead, Ley believes Egede's creature to be similar to another UMO observed a century later in the Gulf of California which resembled a long-necked alligator. Despite mentioning that the size of Egede's ship was not known, he nevertheless estimates that the creature's "large body and long neck" to have reached "to a height of about thirty feet" (Ley 1948:63).

Richard Carrington was a writer of numerous books popularizing science during the middle of the twentieth century. His *Mermaids and Mastodons: A Book of Natural and Unnatural History* is, as the title suggests, an eclectic collection of chapters written in 1957 for the general public about what he calls "mystery animals" from both mythology and paleontology. In the chapter on "The Great Sea Serpent," Carrington rightly highlights the historical importance of the 1734 Greenland sighting, commenting with respect to Reverend Egede that,

> instead of highly coloured variations on a tradition a theme, purveyed at second, third or hundredth hand by simple seamen, we have now for the first time an eyewitness report by a responsible person whose integrity was so far beyond reproach that he was to end his days as an Archbishop (1957:21).

Despite enthusiasm for the truthfulness of the eyewitness, Carrington makes use of the inaccurate version of the illustration from Pontoppidan. He gives no weight to the idea that such creatures might be prehistoric reptiles. And although he states his opinion that Oudemans "puts too much strain upon our credulity," Carrington nevertheless supports the former's contention that the Egede creature could have been a mammal erroneously thought to be extinct:

> Smaller members of the group [i.e., extinct whales called zueglodons] are already known to have survived until at least the beginning of the Miocene Epoch just over 30 million years ago, and there is certainly no reason why some of their larger relations may not even now linger in the seas (1957:24).

The jump in logic in that statement of course makes no sense and is symptomatic of the sort of problematic reasoning that often plagues cryptozoology. An imagined drawing of a zueglodon is offered as a suggestion for sea serpent sightings, including that of Egede.

James Sweeney was a public relations officer with the U.S. Naval Oceanographic Office when, in 1972, he published *A Pictorial History of Sea Monsters and Other Dangerous Marine Life*. It is a richly illustrated coffee-table book designed for a general audience interested in denizens of the deep at a time when Jacques Cousteau's television documentaries were beginning to be broadcast. He examines Hans Egede's comments made about sea monsters contained in his *Natural History of Greenland*, including those about a massive beast called the *hafgufa*, which Sweeney later suggests might have been an emerging volcanic island. He also reflects upon the Reverend's other suppositions about the imagined life cycle and general behavior of sea monsters which are so outlandish as to make one question belief in Egede's reliability in all manners of natural history. Sweeney includes Oudemans' facsimile of Bing's illustration and uncritically reviews Lee's calamari theory. He notes that Egede stated that the "animal reached above our main-top" and that the tail was "an entire ship length from the rest of the body," and follows this by discussing the probable size of round-bottomed merchant ships for that period. From this, Sweeney surmises the head of the UMO to have been more than 20 m above the surface of the water, and the tail to have emerged a distance of 23 m from the rest of the body. His own interpretation of such an improbably gargantuan creature is that it may have actually been,

> the jumpings and cavortings of whales that, in many instances, because of the presence of several, could easily have been mistaken for a single sea monster that performed exactly as Hans Egede described (Sweeny 1972:60).

Richard Ellis is a leading popularizer, writer, and artist of living and extinct marine animals. His *Monsters of the Sea* (1994) is an engaging but unsatisfactory survey of the usual suspect UMOs, such as mermaids, the kraken, the leviathan, Nessie (notwithstanding the book's title), and, of course, sea serpents. The inaccurate "monster and paws" translation of the account is presented, in addition to elsewhere in the book, the blowing water version of the illustration, which is incorrectly identified in the caption as being "Bishop Pontoppidan's sea serpent" (Ellis 1994:186). Confusions continue when later, Ellis presents the fanciful Hamilton illustration, incorrectly attributing it to Hans Egede. Finally, in the concluding chapter, "The New Mythology of Monsters," Ellis echoes Lee's conjecture that not only Egede's creature but a good many other sea monster sightings can be ascribed to the giant squid. It is this dubious conclusion that seems to have stuck in the public consciousness, as it is espoused in the *Wikipedia* entry about Hans Egede, wherein his UMO is stated as "commonly believed to have been a giant squid" (Wikipedia 2021).

In a 2010 article in *Fortean Times*, another magazine popular among cryptozoologists, Mark Greener, a journalist with an interest in anomalous science, reviews what he calls "the golden age of sea serpents." His belief is that misidentified whales account for only a small proportion of sightings, with most remaining "unexplained zoologically, even with the benefit of hindsight" (Greener 2010:1). Greener does not countenance Paxton et al.'s hypothesis of an aroused penis accounting for the Egede sighting. Instead, he contends, based on what he states as "largely" but which in reality turns out to be solely, his trust in the experience of Egede in terms of recognizing whales, that the UMO "is of an otherwise unknown creature" (Greener 2010:2). And so here we arrive, at the end of centuries of varied conjecture, being still, in the minds of some, at a quandary as to what the Egede UMO may have actually been.

Historical Interpretations Synthesized

Twenty-seven of the 36 collated interpretations proposed an identity for the Egede UMO. For these, cryptozoological suppositions outnumbered those positing known animal candidates by a factor of three-to-one, highlighting the importance of this particular sighting in the overall canon of that fringe science. The most popular proposed category (33%) were prehistoric creatures that have escaped extinction: a zueglodon or other archaic whales (8), followed by plesiosaurs and other saurians (3), and a paleo-seal (1). The next most proposed category (31%) consisted of mysterious cryptids that are presently unknown to science: new species of giant reptiles (2) and fish (an eel [1]), and new giant species of mammals (a dolphin [1] and an otter [1]), in addition to traditional bona fide sea serpents (4), and a complete unknown typology (2). Only 22% of putative identifications proposed the UMO to have been an unrecognized existing animal: a giant squid (4), a whale (either several cavorting [1] or mating [1], or a side-view of one [1]), or a giant squid and seal wrestling (1).

What is surprising is that given the long history of recognition that misidentified cetaceans can often be justifiably invoked to explain numerous sea serpent sightings (France 2016c), it was only recently, with the publication of Paxton et al.'s (2005) paper, that the Egede UMO was hypothesized to have been a known species of whale. In contrast, most interpretations proposed candidate animals that are either unknown to science or are generally acknowledged—except by cryptozoologists—to have become extinct. In terms of the latter, it is interesting to see how cryptozoology reflects the faddism of the day, something that even the founder of the field of study himself admitted (Heuvelmans 1968). During the middle of the nineteenth century, the educated public became fascinated with Darwinian evolution, Lyell's concepts of deep geologic time, and in particular, paleontological discoveries such as those of plesiosaurs and mosasaurs (France 2019a; Loxton and Prothero 2015). And sure enough, in keeping with the general trend (Paxton and Naish 2019), this popularity is expressed in three Victorian-era interpretations identifying the Egede creature to have been just such a prehistoric reptile. Then, following the flurry of sightings of the Loch Ness monster during the 1930s, culminating in the now admittedly faked photograph of Nessie (Williams 2015), believers in the continued existence of relict animals escaping extinction (Jylkka 2018) were faced with the difficulty of explaining thriving populations of cold-blooded reptiles in north temperate climates where most of the lacustrine UMO sightings occur. Enter, the new candidate possibility that these freshwater cryptids as well as the Egede UMO observed in the arctic waters off Greenland might have been some form of warm-blooded zeuglodon or prehistoric whale, as witnessed by the eight interpretations suggesting such.

PART II

HISTORICITY AND NEW INTERPRETATIONS

CHAPTER 3.

Agreements and Disagreements with Previous Interpretations, and Proposal of a New Environmental Illation

The previous two chapters in Part I reviewed the course of hypotheses proposed for the enigmatic creature observed that fateful day in July 1734 off the remote coast of Greenland, from that posited by the Egedes themselves to those subsequently made by others over the centuries hence. Table 3.1 summarizes my agreements and disagreements with the various interpretations that have been put forward to explain the Egede UMO. This critical analysis of the historical interpretations arises from decisions made as to the strength or validity of their criticism of preceding conjectures as well as their advancement of novel ones. These assessments were based on my familiarity with the entire corpus of UMO sightings from around the world and through time, as well as experience of four decades of conducting aquatic biology research. The latter attribute is important given that biological interpretations about sea monsters/serpents has largely remained the bailiwick of amateur cryptozoologists who are, according to the comprehensive review by Loxton and Prothero (2015), often woefully unqualified to constructively comment upon the topic.

Disagreements

The Egede UMO should not be ignored (Green, Fabricius, Brown, Heuer). Nor was the UMO entirely different from the imagined Norwegian sea-orm (Pontoppidan) or "Super-otter" (Thomas). Neither was it a prehistoric saurian such as a plesiosaur (Newman, Gosse, C. Gould), a giant squid (Lee, Gibson, Ellis), that might have been wrestling with a large seal (Eggleton and Suckling), a true snake (Landrin), a long-necked paleo-seal (Oudemans), or an unknown reptile (Ley, Eggleton and Suckling). Likewise, neither was the Egede UMO a prehistoric whale, with (Heuvelmans) or without (Carrington, Thomas x 2, Hynes) super-otter like traits, or possessing vestigial (Eberhart) or completely formed (Shuker) limbs. Nor was it a true giant otter (Woodley), a giant dolphin (Woods), a giant eel (Burton), a group

Table 3.1. Agreements and disagreements with interpretations advanced concerning the mysterious case of the Egede UMO (Unidentified Marine Object).

SOURCE	AGREEMENT	DISAGREEMENT
1743 Green		Ignores as unworthy of mention
1755 Pontoppidan		"Sea serpent" different from the Norwegian sea-orm
1767 Cranzt	Similar to the Norwegian sea-orm	Restricted to Nordic regions
1780 Fabricius		Ignores as unworthy of mention
1810 Ball/Coleridge (cited in Heuvelmans [1968])	Not just endemic to Greenland	
1839 Hamilton		Exaggerated account
1849 Newman	Not a sea serpent or snake of any kind	Prehistoric plesiosaur
1860 Gosse		Ignores since doesn't fit theory of prehistoric saurian
1868 Brown		Dismisses as mere fantasy
1875 Landrin	Similar to the sea-orm	Marine reptile—a true snake
1883 Lee		Giant squid
1884 Woods		New species of giant dolphin
1886 Gould, C.	Nothing to do with serpents Similar to other sightings elsewhere	Possible prehistoric saurian
1887 Gibson		Sea serpents exist Giant squid theory credible
1890 Ashton	Somewhat similar to the sea-orm	
2007 [1892] Oudemans	Giant squid theory not tenable	Long-necked paelo-seal
1902 Hoyle		Giant squid Possible existence of sea serpents
1930 Gould, R.	Text more reliable than illustration Not a giant squid	Cannot have been a whale or shark
1948 Ley	Reliable account. Body and neck raised about 10 m high	Not a whale but an alligator-like animal. Nothing in common with the Nordic sea-orm
1954 Burton	Most sea serpent accounts can be explained	Probably a giant eel
1957 Carrington	Not a prehistoric reptile	Possibly a zeuglodon
1968 Heuvelmans	Exhalation suggests not a whale	Super-otter-like zeuglodon 30 m long
1972 Sweeney	Tail 23 m from body	Body 28 m above water Group of cavorting whales
1994 Ellis	Giant squid	
1996 Shuker		Oldest archaic whale, with four large limbs and looking like an otter Possible zeuglodon
1996 Thomas	Protuberance a possible dorsal fin Barnacle encrusted skin like a whale	Different from Norwegian super-otter UMO Possible zeuglodon
1998 Eggleton and Suckling		Possible large seal wrestling with a squid Unknown reptilian "monster"

Table 3.1 continued.

SOURCE	AGREEMENT	DISAGREEMENT
2003 Coelman and Huyghe	Ignores Heuvelmans' super-otter	Lumped into broad classic sea category serpent category which might be zeuglodons
2002 Eberhart		More primitive whale than the basilosaurids, and having vestigial limbs
2005 Paxton et al.	Possibly a known species of whale Size of 42–58 m thought realistic Illustration not wholly accurate Body and tail not raised simultaneously in air	64–98 m size believed to be unrealistically long Possibility that presumed tail is missing flukes Possibility that presumed tail is an aroused penis
2008 Woodley	Penis theory doesn't measure up	True giant otter
2010 Greener	Not an aroused penis	Whale explanation wrong Creature unknown to science
2011 Thomas	Egedes would have recognized a whale Dorsal fin shown is an important identifying trait	Zueglodon
2010 Drinnon	Drawing is of some value	Not an aroused penis but possibly an oblique view of a normal tail of a grey whale
2012 Hynes		Complete dismissal of known whale possibility Prehistoric cetacean
2019 Heuer		Ignores as unworthy of mention

of cavorting whales (Sweeney), or some unknown animal in general (Greener). However, neither can the UMO be completely precluded from having been a whale (R. Gould, Greener, Hynes), or a shark (R. Gould), thereby leading to it being categorized as a "classic sea serpent" (Coleman and Huyghe) based on the incongruity of having a serpentine extension that is presumed to be the tail, but which is unlikely to be due to missing flukes or a misidentified aroused penis (Paxton et al.). The account was not greatly exaggerated (Hamilton), and the body of the UMO was not fully 28 m above the water (Sweeney). The overall length estimates of 30 m (Heuvelmans) are too short, whereas there is no cause to dismiss derivations of a total length in the range of 64 to 98 m as being unrealistic (Paxton et al.) for the simple reason proposed below that the presumed tail is *not* biological in nature. Finally, and perhaps most controversially in the light of modern interpretations, the assumption that the animal was (Paxton et al.), or at least *has* to have been, a cetacean can be challenged; i.e., it is not necessary to conclude that the Egede UMO was a marine mammal that originally went unrecognized at the time of the sighting (Heuvelmans) but which today should be identified as a known species of whale (Paxton et al.), whose skin is covered with barnacles (Thomas) and whose tail fins are hidden (Drinnon).

Agreements

The UMO observed by Egede is of a type that is found beyond Greenland (Ball/Coleridge, C. Gould), and although it has traits that resemble the Norwegian sea-orm (Cranzt, Landrin, Ashton), it need not be placed in an invented category of super-otter (Coleman and Huyghe) or that of any type of sea snake (Newman, C. Gould). It is neither plausible that the UMO was a giant squid (Oudemans) or a prehistoric reptile (Carrington). Most sea serpent accounts can be explained without having to invoke cryptozoological fantasy (Burton), as, for example, the good possibility that the UMO was a species of whale (Paxton et al.). Although the original text is likely more accurate than the accompanying illustration (R. Gould), the latter is useful (Ley, Drinnon) in showing a dorsal fin (Thomas x 2) and in supporting the text in disqualifying, based on size, the possibility that the presumed tail, which contributes to the immense overall length of the UMO in the range of at least 42 m and possibly as much as 50 or 60 m based on calculations made in relation to ship sizes (Sweeney; Paxton et al.), might have been an aroused penis (Woodley, Greener). That said, both the body and the so-called tail could not have been simultaneously elevated above the water (Paxton et al.), as the illustration depicts.

New Environmental Interpretation: Non-Lethal Entanglement

It is my contention that the animal observed by Egede leaping out of the water was a known but unfamiliar species, one that in this case was affected by anthropogenic activity. And so then, what could this strange being, this hybrid animal with strong whale-like traits but sporting an incongruous, elongated and fluke-less tail, have possibly been? For the answer to that question, we have to leave the cryptozoology world of imagined animals behind and instead enter the conservation biology world of real environmental problems (France 2017, 2019a, 2020a).

One of the few things that many of the commentators, pseudo- and real scientists alike, agree upon, is that the only serpent-like feature of the Egede UMO was its lengthy tail. For example, from Heuvelmans (1968:102): "While Olaus Magnus's animal [i.e., that shown on his famous sixteenth-century map] was serpentine in its head and neck, only the tail is so in Egede's monster"; and from Paxton et al. (2005:3): "It is only the rear end/underside of the monster that is described as serpent-like." Many sightings of UMOs describe them as being strange chimera creatures. Eyewitnesses often first observe something they recognize as a known animal, as for example a whale or a sea turtle. It is only later, upon spotting the long extension which they logically presume to be the creature's enormous tail, that they shift their overall opinion to that of the creature being something unusual. The only way they can explain the anomaly is to imagine that what they have observed is some type of hitherto unknown animal. And thus, from such head-scratching encounters, ergo are born sea monsters or sea serpents. From descriptions and particularly illustrations of such UMOs (as, for example, those shown in Figure 3.1) it is obvious that these particular UMOs are actually animals pulling a string of anthropogenic material behind them (France 2016b, 2016c, 2017, 2018, 2019a, 2019b, 2019c, 2020a, 2020b). It is therefore my belief that Egede's "most dreadful monster" can be parsimoniously explained as being one of the earliest described examples of an animal observed entangled in active fishing/hunting gear or the flotsam/jetsam of some casually abandoned (e.g., drifting "ghost" nets) or deliberately discarded maritime debris.

Today, entanglement is recognized to be a serious environmental problem occurring on a global scale (Gregory 2009), and one that notably affects many species (Laist 1997). It is important to note that entanglement need not result in immediate death, as there is ample evidence that cetaceans can pull trailing anthropogenic debris for considerable periods following initial entanglement (Johnson 2005). For example, a whale pulling entangled debris, and thereby christened "Necklace," was repeatedly observed over the period of years during her annual migration into the Bay of Fundy (Fama 2012). Other entangled whales are known to have traveled thousands of kilometers (BBC News 2019; Brown 2019; Lyman in NOAA 2014).

The current belief is that entanglement originated in modern times with the advent of non-degradable plastic in the middle of the twentieth century (e.g., NOAA 2014; Vegter et al. 2014; Wabnitz and Nichols 2010). This is based on the assumption that material made from natural fibers such as hemp "will lose their resilience in usage and if lost or discarded at sea [will] tend to disintegrate quickly" (Gregory 2009:2014). Documented cases of entangled whales only date back to the late 1980s, and the early 1990s for many other animals (Gregory 2009). However, it seems that this might be more a consequence of the unrecognition and absence of studying the problem by conservation biologists before that time, rather than the absence of the existence of the actual problem itself.

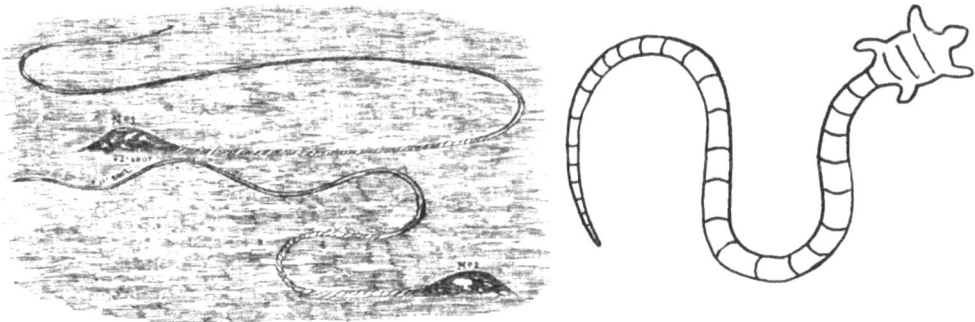

Figure 3.1. Mystery or mundanity? Cryptozoology or conservation biology? Two sea monsters completely new to the scientific knowledge of existing biodiversity and the fossil record, or simply (and parsimoniously) early examples of the all-too-common phenomena of a non-lethally entangled cetacean (left panel; South Africa 1857; note the "tail" resembling a braided rope) and a non-lethally entangled chelonian (right panel; China 1925; described as a "turtle-bodied snake"; France 2016b:9, 2018:112)?

Although it is true that compared to modern ropes made of synthetic fiber, ropes and nets made from hemp, flax, or cotton in the eighteenth century will have deteriorated more rapidly, it would be erroneous to suppose that such material would decompose quickly given its continued widespread maritime use for millennia (Aiken and Purser 1936; McCaskill 2009). As such, pre-plastic fishing material would have lasted long enough to pose a threat for entangling susceptible animals. For example, a careful parsing of the words and examination of the accompanying illustrations in accounts of what were thought to have been sea serpents/monsters with long sinuous extensions that were presumed to be their tails (France 2016c, 2018), or those grabbing onto or wrapping themselves around whales during presumed ferocious battles (France 2016c), suggests that entanglement has probably occurred for as long as humans have been placing equipment in the oceans. Experts interviewed by Deedy (2017) agree that anything left in the ocean has always presented an entanglement hazard for marine life, regardless of whether it was made of plastic or natural fiber.

Fauna in European and North American coastal waters were certainly experiencing severe, and in some cases, non-sustainable, fisheries exploitation at the time of the Egede sighting in the middle of the eighteenth century (e.g., Kittinger et al. 2015). Given the seasonal distribution patterns of whales and large fishes in the North Atlantic, it is entirely possible that the animal seen by Egede off the coast of Greenland could have transported a train of fishing nets and floats (Figure 3.2) that had their origin on either side of the ocean where the entanglement had initially taken place. Floats at the time were made of blown-glass balls, pieces of cork, or even barrels and casks (photographs of such fishing gear are presented in France 2019a, 2020b). Interestingly, it is just this sort of recurring seasonal appearance and disappearance that has been posited to explain the UMO seen for about a month in each of several successive years in the early nineteenth century in and around Gloucester, Massachusetts (Fama 2012). This Gloucester UMO remains the most sighted and studied sea serpent in history, and is almost certainly to have been an animal entangled in fishing gear (France 2019a, 2019b). It is also entirely possible that the train of entangled fishing gear originated within the Davis Strait itself, given that Europeans at the time were also traveling to Greenland to harvest the rich fish stocks (Figure 3.3). Furthermore, as hemp rope would often be impregnated with dye or pine tar to slow down the rate of deterioration, thereby enabling it to last for many months, the trailing anthropogenic debris could have itself become entangled with long strands of kelp. Today, animals that are festooned with entangled mats of kelp certainly do give the impression of having solid tails.

Could fishing debris have been what the UMO trailed when its elongated or serpentine "tail" was observed by Poul Egede and Reverend Bing? After all, Norwegians have used nets since the Middle Ages to block fjords to entrap whales. Additionally, Magnus, in his *Historia*, mentions that "sea-monsters or whales have been hauled out of the sea" (in Szabo 2008:86), and Teit (1918) recounts that whales ruining nets were considered to be monstrous sea-trolls that were in demonic possession of the hapless cetaceans. And in what is one of the earliest

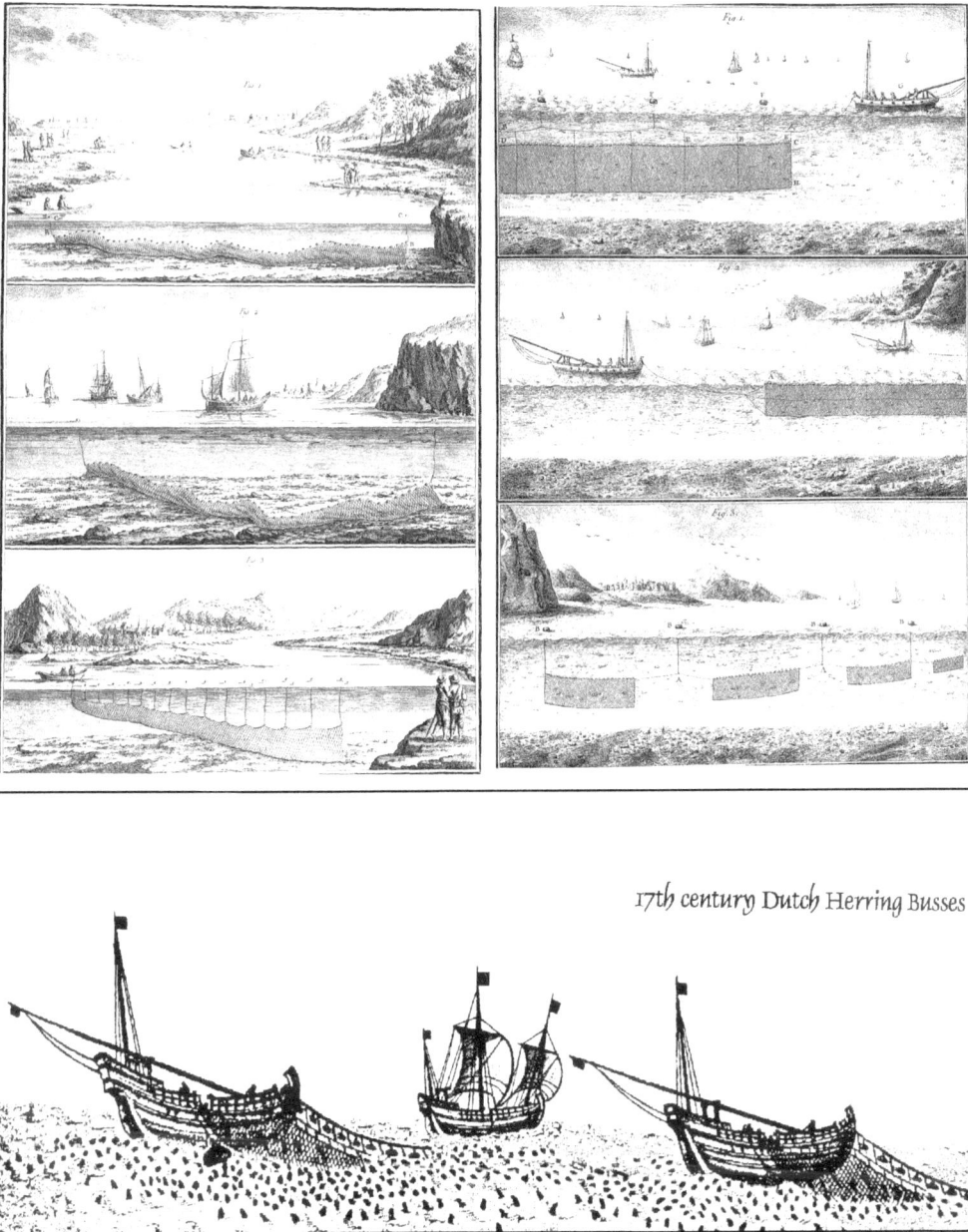

Figure 3.2. Strings of fishing net-floats of the types used at the time of the Egede sighting which may have comprised the "tail" entangled about the UMO. Top-right triplet of illustrations shows deployments of surface, mid-depth, and bottom nets, the latter two using the two-tiered system of buoyancy for a "sunk net," including a surface warp-line of casks and a submerged net-rope of smaller cork pieces (Duhamel du Monceau 1769:235). Top-left triplet of images shows fishers working with nets which have strings of surface floats (Duhamel du Monceau 1769:318). Bottom image shows the seventeenth-century Dutch herring fishery, from an unspecified source in Butcher with his caption (2000:59) stating: "the wooden barrels used to float up the gear are clearly visible."

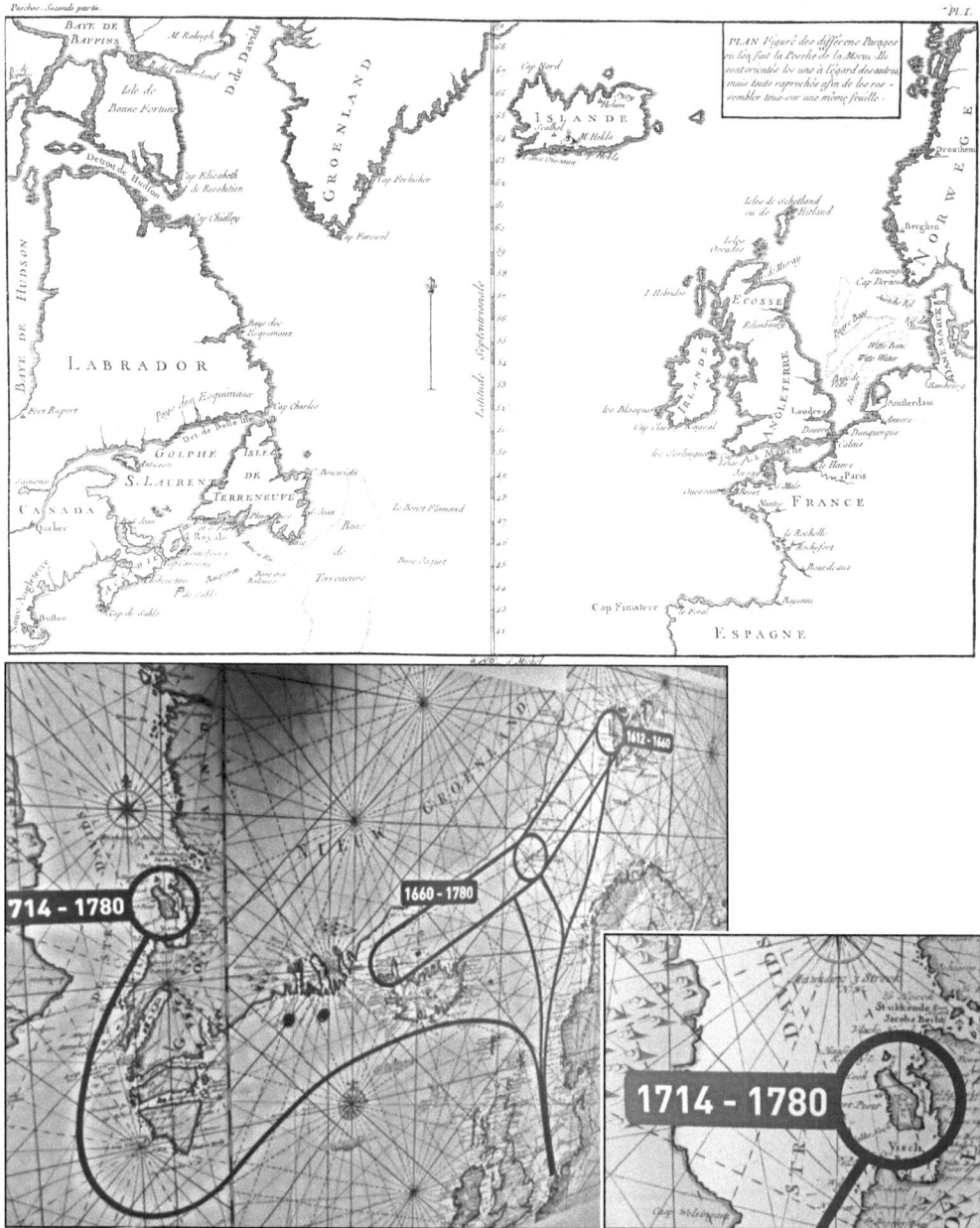

Figure 3.3. Top image: Map from Duhamel du Monceau's seminal book (1769:68) about fishing in the North Atlantic by Europeans during the eighteenth century, showing the fishing ground in the Davis Strait between Greenland and Baffin Island, contemporaneous to the Egede sighting. Lower two images: Photographs of a wall poster in the Amsterdam Maritime Museum of an enlargement made from Reiner and Joshua Ottens' 1740 hand-colored chart of Greenland and the Davis Strait. Modern additions show the westward progression of Dutch whaling activities, from Spitsbergen in 1612–1660, to Iceland and the east coast of Greenland in 1660–1780, and to the west coast of Greenland in 1714–1780. The latter is significant, as shown in the close-up, because it is centered around Disko Island in Baffin Bay, near where Poul Egede and Reverend Bing sighted their UMO in 1734, at a time when whaling was at its peak and the likelihood of struck animals escaping capture by swimming off with a train of entangled ropes and floats was the greatest.

mentions of "bycatch," Magnus refers to "great monstrous creatures being caught accidently in nets or on hooks" (in Szabo 2008:203), before going on to describe one such monster that sported a beard-like arrangement of long, thick hairs that resembled goose feathers. This, Szabo conjectures, might have been baleen plates, but the "hairs" could alternatively have been the strands and mesh of entangled fishing nets. The sea serpent literature is filled with numerous accounts of UMOs identified as having long "manes" draped about their backs, including the famous *Daedalus* encounter (Galbreath 2015), which has likewise been postulated to have been an entangled animal (de Camp and de Camp 1985). It is possible that the presumed tail of the Egede UMO was indeed comprised of such entangled fishing net material that originated from a distant location. But it is also possible that there is another related and just-as, or perhaps even more likely, potential explanation, particularly so given that the UMO sighting took place in a region where whales were regularly hunted (Figure 3.3).

During the eighteenth century, whales were being decimated at alarming rates through direct hunting pressure, leading, in the case of the grey whale, to their complete extirpation from the north Atlantic. Harpoon hunting of basking sharks began in earnest at the same time. The hunting technique that was most frequently used in both cases is referred to as "kegging." Here, a harpoon, attached to large floats such as barrels or sometimes even a small dory, is stuck into the prey (whales, sharks, giant tuna, marlins, etc.) to create increased resistance to slow down the animal, to track its movements for hours and occasionally for days, to ensure that the struck animal remained buoyant after its death, and finally to facilitate towing its carcass to shore (photographs of such hunting equipment are presented in France 2019a, 2020b). Magnus, for example, describes harpoon hunting one of the "monsters" shown on his map, a *xiphias* (likely a swordfish), that was finally retrieved after exhausting itself in such a manner. Szabo (2008) mentions that for medieval fishermen, the tracking and towing of a struck whale was a more difficult task than its killing, and that the floats attached to spear-lines were made from inflated seal or cetacean stomachs. J. Tuck and R. Grenier (in Szabo 2008:112) describe the use of inflated drags by sixteenth-century Basque whalers off the coast of Labrador:

> [A] whale is harpooned with a barbed iron harpoon attached to a sturdy oak shaft. A 'drogue' or drag, attached by a rope to the harpoon was thrown overboard from the whaleboat. Towing the drogue slowed and tired the whale and as it surfaced to breathe, additional harpoons and drogues were made fast.

Moreover, marine megafauna, notably whales, have been hunted by Inuit in the Eastern Arctic for at least a millennia (e.g., Douglas et al. 2004; MacNeil et al. 2012; Whitridge 1999), employing similar techniques of hurling harpoons attached to lines of floats and drag-gear from unimak boats (McCartney 1980; Taylor 1979). Cranz (Crantz 1767:120) provides the first European description of Greenlanders' whale-fishery:

> They assail the whale courageously in their boats and kajaks [sic], dart-
> ing numerous harpoons into his body. The large seal-skin bladders, tied to
> these weapons prevent him from sinking deep into the water. As soon as he
> is tired out they dispatch him with short lances.

Consider next, the situation which Magnus stated occurred frequently, wherein the struck animal did *not* succumb but instead got away or the ropes having to be deliberately cut from the boat if the whale swam off on the surface rather than dove straight down. This, I believe, is the stuff from which dreams of sea monsters are made of, to paraphrase the closing line of Shakespeare's from *The Maltese Falcon*, a film (and book) about another famous and actively sought for mythological animal. Harpooned cetaceans are known to be able to survive for extended periods (BBC News 2019; Brown 2019; Gardner 2007), and in pre-ballistic times, it is thought that as many as one-quarter of all struck whales actually managed to evade capture (Mowat 1997). For example, noted polar explorer Fridtjof Nansen, wrote of the frustration arising from such an encounter:

> I felt pretty sore at losing such a good boat. Well I wasn't going to risk an-
> other boat, but I thought I'd be even with him all the same. Next year I
> took some petroleum casks with me. I rigged up three of these casks, fixing
> them on to three new whale-lines, and laid them all ready at the bottom
> of the boat. Then we started out again. Well, I made fast to a fish [i.e. a
> whale], and down he went in the same way. The first line ran right out and
> we chucked the first cask overboard. But he pulled it on down with him
> full pelt as before, without stopping a moment.... Then the second line ran
> out, and we chucked the second cask into the water, but it just went after
> the first and disappeared in the same non-stop fashion; while the third line
> went running out as fast as though we hadn't had any casks at all. At length
> we chucked the last cask overboard; but I'll be handed if it didn't go under
> every bit as quick as the others. So we'd lost the whale and the lines and
> the casks, and we never saw them again (in Mowat 1997:299 and France
> 2016c:28).

It is therefore entirely conceivable, and not at all surprising, that should Egede and Bing have spotted a similarly struck and escaped whale or basking shark that was engaged in the normal behavior of breaching, to which were attached intertwined ropes with floats or casks, possibly also augmented with entangled fishing nets or festooned with kelp strands picked up along the way, they could easily have imagined the entity to be a "very horrible sea-creature" of unknown variety. And this is what Pontoppidan would later transform into being a mon-strous sea serpent, thereby birthing the field of aquatic cryptozoology.

One further point of evidence precludes the long extension of the Egede UMO, which all previous commentators with the exception of Paxton et al. (2005) have unquestionably presumed to have been a tail, from being an actual body part of the creature. Look again at the various illustrations of the UMO, and how it is portrayed in a tightly bent U-shape with both the front-end and tail simultaneously raised up high into the air. This is of course a physical impossibility, which could only occur had the animal been dropped onto the surface of the water from a considerable height (there are "para-cryptozoologists" whom actually *do* believe that sea serpents and lake monsters have extraterrestrial origins from flying saucers). In a similar vein, many have noted over the years that accounts of sea serpents with coils of their tails elevated above the water surface are a non-starter due to invalidating basic laws of biomechanics. But again, this is where Bing's illustration (in all its various renditions) is shown to be a less reliable source than the original textual account. If we carefully read that encounter again: "It was otherwise created at the rear like a serpent and when it went under the water it lifted itself backwards and raised then the tail up from the water a ship's length away from the body" (H. Egede in Paxton et al. 2005:2), we can see the error in the illustration. The reality is that it was only when the front-end of the UMO had plunged backwards into the water after executing what we now recognize to be a straightforward display of breaching behavior, *then* the extension of the presumed tail was lifted upwards into the air. Paxton et al. (2005:3) are therefore correct in stating: "The animal had a serpent-like tail that appeared out of the water when the rest of the beast had disappeared," the confusion due to the illustration not being "wholly accurate."

One of the 13 physical and behavioral attributes of the purported Gloucester Sea Serpent, which was almost certainly to have been a misidentified entangled marine animal (Fama 2012; France 2019a), is "Extended body pulled down into water, thrown up into air, or thrashed about on the surface." For 27 sightings of different sea serpents from around the world and over the centuries that I have proposed to have been entangled whales, this particular attribute was the third most prevalent trait displayed; following "Narrow, tapering, sinuous, snake/eel-like shape, often with absence of a tail fin or lateral appendages" and "Notable/unusual length" (see France 2016c). All three of these traits clearly pertain to the Egede UMO, with the account descriptions reading similarly.

The first time I saw a breaching whale was during an ecotourism cruise in the Stellwagen Bank National Marine Sanctuary, off the coast of Massachusetts. Unfortunately, the magic of moment was quickly dashed when, as the humpback whale completed its backward plunge, it was shown to sport a long "tail" that, due to our close proximity, could be discerned to be an entangled string of herring net floats thrown up into the air. And just like flicking a giant skipping rope, the entangled net material moved in an animated and twisty fashion in consequence of the momentum of the whale's plunge back into the water. Should our boat have been farther away from the breaching animal or the weather conditions have been less ideal, and should we have been in ignorance of the modern-day prevalence of entanglement, there is no doubt that the impression left would have been quite different from the reality. Even so, as it was, one excited tourist shouted out, "look at its tail!" before being informed otherwise by the knowledgeable guide. Likewise, the literature is rife with examples of how people have been fooled into believing that they have spotted an elusive sea serpent/monster by long strings of nets attached to docks which bob up and down in tidal currents or by floating mats of seaweed that undulate through wind or wave activity (see examples in France 2018, 2019a, 2020a).

When an entangled animal, misconstrued by observers as a sinuous sea serpent, plunges into the water, it can throw its length of trailing debris up into the air. This is what I observed off Cape Cod, and this was probably the case for the famous 1890 sighting of the "Moha-moha" UMO in Australia (France 2017), which many have surmised over the years to have been some hitherto mysterious turtle-*like* animal, but which was likely to have really been a chelonian, but one that was entangled (Figure 3.4). Consider also the 1879 sighting of a whale being attacked by a purported sea serpent that was observed by the crew of the *Kirushiu-maru*, off the southern coast of Japan. Of the hundreds of sea serpent sightings, it is this one that may come closest to resembling the Egede encounter. The text describes the erroneously presumed battle:

> The chief officer and myself observed a whale jump clear out of the sea, about a quarter a mile away. Shortly after it leaped out again, when I saw there was something attached to it. Got glasses, and on the next leap distinctly saw something holding on the belly of the whale. The latter gave one more spring clear of the water, and myself and chief then observed what appeared to be a large creature of the snake species rear itself about thirty feet out of the water. It appeared to be about the thickness of a junk's mast and after standing about ten seconds in an erect position, it descended into the water, the upper end going first. With my glasses I made out the colour of the beast to resemble that of a pilot fish (France 2016c:15).

Again, note the temporal sequence here that parallels the Egede encounter; i.e., the creature first breaches and plunges down, "then" the serpent appears "rear[ing] itself" 10 m into the

Figure 3.4. The Moha-moha, or "monster turtle fish," observed in Australia in 1890 (Heuvelmans 1968:295), uplifting its "tail," which was described earlier, when the UMO was observed on the beach, as being jointed and shiny (which reads like a string of glass floats used on fishing nets— France 2017).

air. Most interesting are the two illustrations that accompanied the publication of the encounter (Figure 3.5). The left drawing shows a typical breaching whale with a thick line of debris entangled over its right fin, whereas the right drawing shows the same whale moments later, heaving the entangled gear or debris up into the air, with tail flukes raised upward during its plunge back into the water. Compare this latter drawing to Figure 1.7, depicting what I believe to be a train of entangled debris of the Egede UMO lingering momentarily in the air after the creature has plunged back into the water. The only real difference is the absence of a bilobate tail seen for Egede's entangled animal, since the body was already beneath the surface at the precise moment being depicted.

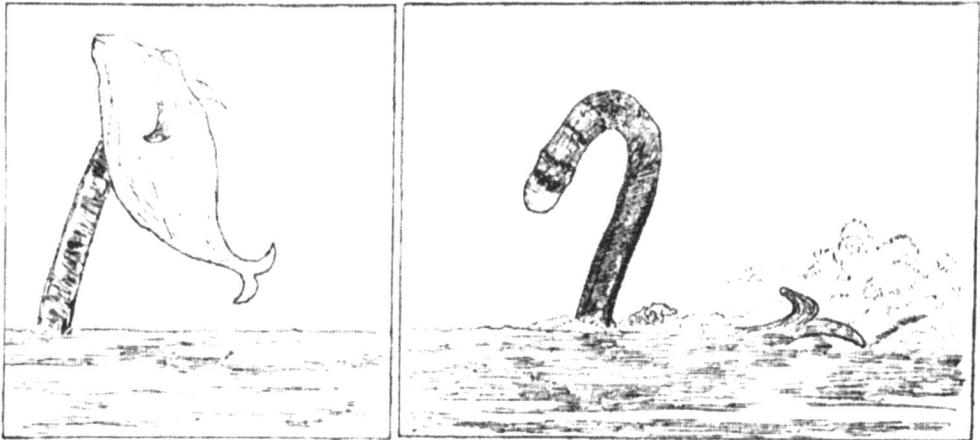

Figure 3.5. Depictions of a purported sea serpent attacking a whale off the coast of Japan in 1879 that resembles (left drawing) a normal breaching whale with a thick line of debris entangled over its right pectoral fin, and the same whale moments later heaving what I contend to be entangled fishing or hunting gear upward into the air while raising its tail to plunge back into the water (France 2016c:16). The likeness to the Egede encounter is quite marked.

The Long Tail: Penis or Peril?

The question comes down to whether the serpentine extension observed for the Egede UMO is evidence of an aroused animal displaying a penis or rather of an entangled individual trailing some type of anthropogenic debris. Paxton et al. (2005:8) write that

> our explanation also assumes that the witnesses would not have recognized a whale's penis and that some species would display their penises in the summer off Greenland. Hans Egede (1741, 1745) described the large 'membrum virile' of a whale but the Egedes may not have realized it could be seen at sea.

But this seems difficult to accept for individuals who were so knowledgeable about natural history. Hans Egede (1745:181), himself, states:

> The penis of a whale is a strong sinew, seven or eight, and sometimes fourteen feet long, in proportion to his bulk: it is covered with a sheath, in which it lies hidden, so that you see but little of it.

He then goes on to describe copulation in the water, including minute details such as the consistency of whale semen. Certainly he must have realized, as did others at the time, that a distended penis, as often observed in dead whales on the shore (Figure 3.6; there is a humorous online example of an overly enthusiastic cetacean biologist waving about just such a member while giving an impromptu lecture), would be used in copulation at sea, and that something of such prodigious proportion could occasionally be visible during the process. When one gets over the initial shock of seeing a "dropped" penis of a stallion in a paddock, the so-called "fifth-leg" due to its size, no one would be at all surprised to see it used should observations be made of mating horses in a field. There is a logical consistency of equating either a shore- or paddock-observed penis to ocean or field copulation. Furthermore, the Egedes' descriptions make no mention of the color or surface texture of the UMO's "tail." It seems unlikely that they would have not recognized an aroused whale penis, especially given that the organ's light color (thereby leading scientists to refer to them as "pink floyds") is so distinctive and different from the rest of the generally dull, green-grey or black coloration of the rest of the body of most large cetaceans.

In the cases noted above for what may have been entangled animals misidentified as sea monsters, the considerable length noted for the serpentine tail means that they obviously could not have been aroused penises, which would have been of a substantially miniscule size relative to that reported and illustrated. Others have likewise noted the size-disparity conundrum and in consequence have not been accepting of the "cetaceans, sex and sea serpents" theory, despite its cute alliteration (Greener 2010; Thomas 2011; Woodley 2008). Paxton et al. (2005:3) get around this shortcoming in their theory by suggesting "the presence of more than one male whale" who were "at one point a ship's length from the body." This explanation, requiring as it does the presence of now not one but rather three or possibly more whales, all of a species that collectively went unrecognized, is not as parsimonious an explanation as is that of the UMO being a single animal which had become entangled. Likewise, Paxton et al.'s aroused penis explanation does not stand up for their explanation of the 1875 *Pauline* encounter of a pod of sperm whales that were described as being "frantic with excitement," one of which was believed to in the process of being attacked by "a monstrous sea-serpent coiled twice round a large sperm whale" (complete text of the account in France 2016c:17). Again, the 10 m estimated length for the titular sea serpent in that case is much longer than that of any whale penis, suggesting entanglement to be the more likely explanation for the sighting. However,

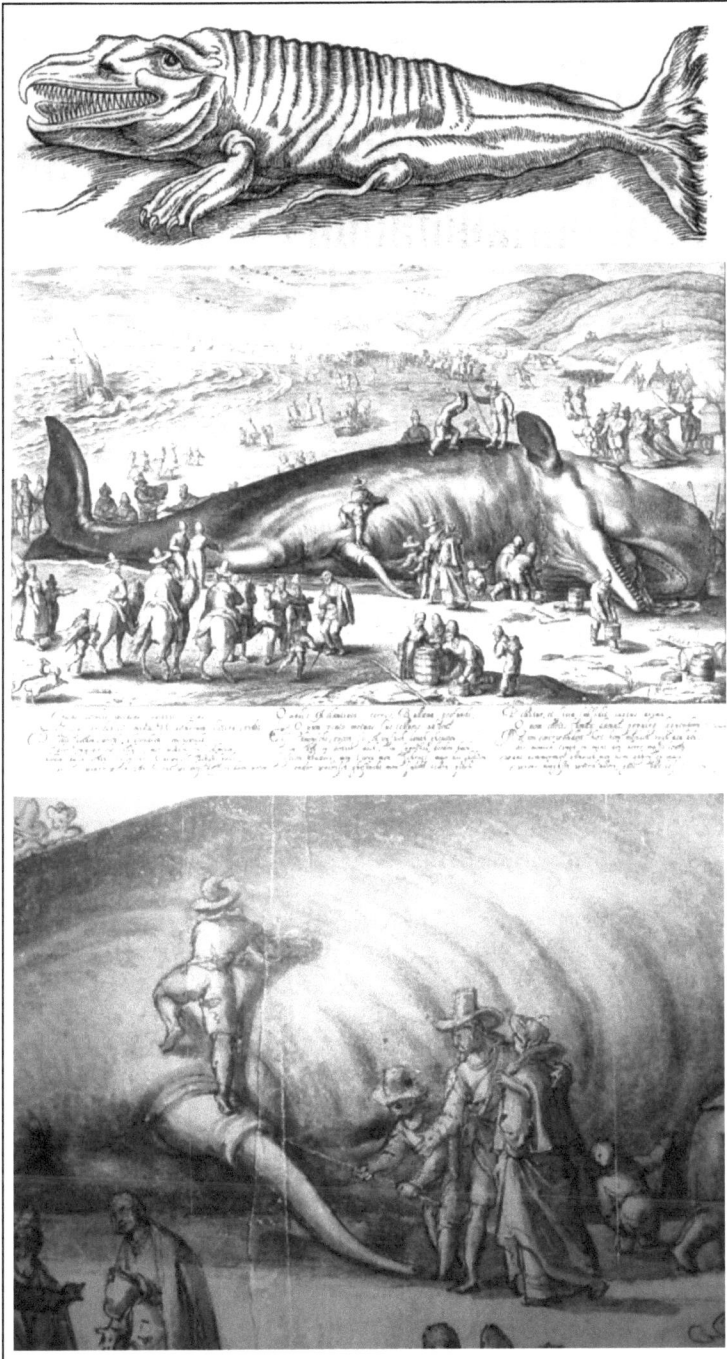

Figure 3.6. The upper panel shows a stranding of a monstrous whale, from Conrad Gessner's sixteenth-century classic *Historia Animalium* (1551–1558:621), consided the birth of modern zoology. Note the extended penis, a common occurrence in such instances. The middle panel is Hendrick Goltzius' 1598 illustration of people viewing a beached sperm whale in the Netherlands. The lower panel is a close-up, cropped from this image. Note the natural historian who is engaged in measuring the prodigious length of the unfortunate animal's penis. Middle and lower panel photo courtesy of Wikimedia Commons. See Photo Credits page.

it is possible that in the *Pauline* case, what was believed to have been a single sea serpent in battle with the whale might have been the prehensile penises of *several* amorous males curving around the female's belly, presented in lordosis behavior in preparation for copulation, each phallus endeavoring to be the first to reach the vaginal opening of the receptive female (i.e., unlike for the Egede encounter, eyewitnesses of the *Pauline* incident *did* record the presence of more than one large animal, all of which, it should be noted, were clearly recognized to have been whales). One can view films of just such orgy-like breeding encounters of cetaceans online. Additionally, and in contrast to the Egede UMO, that of the *Pauline* is described as being a "white pillar," which does give credence to Paxton et al.'s illation for this particular case. Still, given the accompanying illustration (Figure 3.7) and description of the *Pauline* account, entanglement can be thought of as being just-as, or even the more likely, explanation.

That said, there are other sightings of purported sea monsters whose "heads" were observed elevated just a few meters above the surface of the water, and which do resemble the famous but now discredited fraudulent Nessie-as-a-plesiosaur photograph, that are much easier to imagine as being aroused penises than either the *Pauline* or the Egede encounters (Figure 3.8). As is also the case with regard to the following description of an UMO from Oceania that was seen in 1923:

> There was considerable movement in the sea. I saw a strange animal raise its head, neck and beginning of its body out of the water and stand straight up like an erect snake and then fall back, hitting the sea, and raising a great

Figure 3.7. Penis or perdition? The 1875 *Pauline* sighting of a cetacean believed to be engaged in mortal combat with a purported sea serpent unknown to science (Oudemans 2007[1892]), or simply a sperm whale encircled twice around by either a single aroused and enormous penis or possibly several such (in either case from males that remain hidden from sight; Paxton et al. 2005), or a typical train of fishing or hunting gear wrapped around the entangled animal (France 2016c:16)?

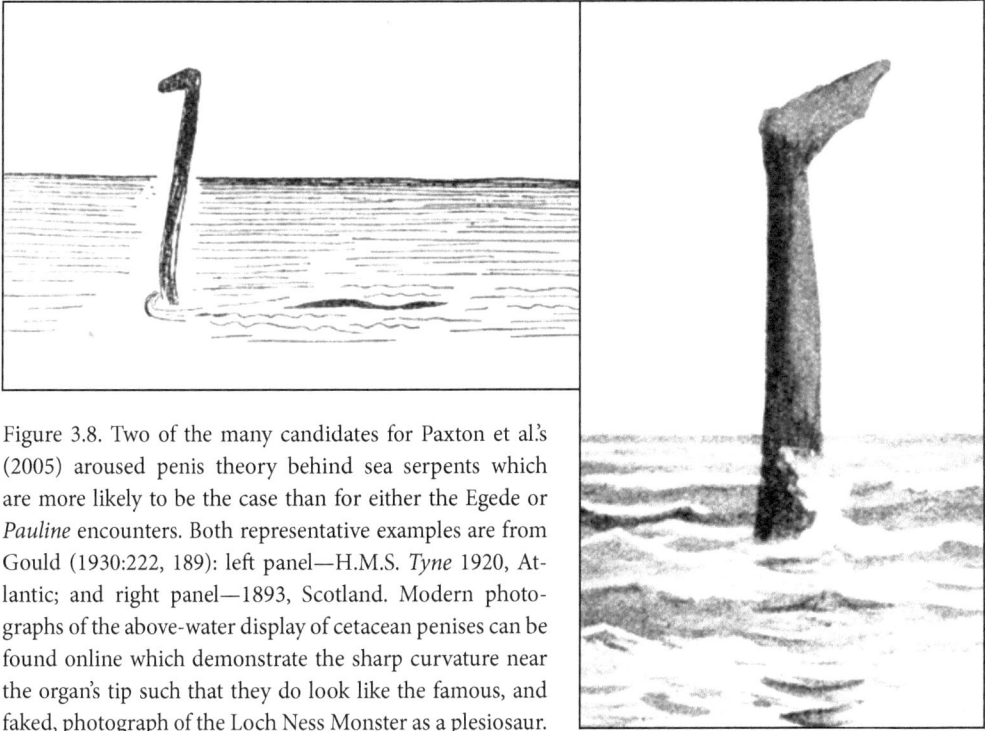

Figure 3.8. Two of the many candidates for Paxton et al.'s (2005) aroused penis theory behind sea serpents which are more likely to be the case than for either the Egede or *Pauline* encounters. Both representative examples are from Gould (1930:222, 189): left panel—H.M.S. *Tyne* 1920, Atlantic; and right panel—1893, Scotland. Modern photographs of the above-water display of cetacean penises can be found online which demonstrate the sharp curvature near the organ's tip such that they do look like the famous, and faked, photograph of the Loch Ness Monster as a plesiosaur.

> plume of water... It emerged a second time... but this time did not dive with a splash but sank slowly down (Heuvelmans 1968:414).

Other descriptions of this particular UMO's extension state that it was of the proportion of a boiler-room air-shaft, was slightly pointed at the end, and was light and "rather dirty" in colour; all of which certainly do suggest it could have been an aroused cetacean penis.

As the humorous adage has it, when considering the alternative theories for these UMO extensions, size does indeed matter. Reverend Bing's illustration portrays the supposed tail and body of the mystery animal as being in close proximity. But this U-shaped depiction of the entity is inaccurate when one reexamines the original text. There, it is clearly stated that the "tail" was raised up from the water fully "a ship's length away from the body." In other words, the two structural components, body and tail, were separated by a considerable distance between them. If one cedes authority to Hans Egede's capabilities as an accurate describer of natural history—as most commentators have when they consider the demonstrated reliability of his entire corpus of writings—then it is impossible to reconcile that the tail as so described could have been a penis. So whereas Paxton et al. could be, and probably are, correct that a proportion of all sea monster sightings may be explained this way, it is doubtful that aroused penises are the most likely explanations for the particular cases of either the Egede or the *Pauline* UMOs. As a result, Hans and Poul Egedes' descriptions and Reverend Bing's illustration may be of environmental historical significance in that they represent one of the earliest cases of an animal observed to be entangled in anthropogenic material. In particular, the Egede UMO, spotted that day off the remote coast of Greenland, may have been an animal fortunate to have escaped capture but one left pulling a trailing line of floats attached to a harpoon embedded in its flesh as a legacy of that near fatal encounter with the rapacious eighteenth-century whaling or shark hunting industries (or possible local Indigenous harvesting).

But what then is to be made of the following two sightings of UMOs, whose long extensions were estimated to be of sizes intermediate between those of cetacean penises and the mysterious Egede "tail"? The first occurred in Scotland in 1900 and was referred to by Heuvelmans as another example of a "long-necked sea-serpent." While hauling in a net the eyewitness observed,

> an object rising out of the water about fifty yards to seaward of them. It was about a yard high when he first saw it, but, as he watched, it rose slowly from the surface to a height of twenty or more feet—a tapering column that moved to and fro in the air … While this 'tail' [eyewitnesses' use of quotation marks] was still moving in the air they could see the water rippling against a dark mass below it which was just breaking the surface, and which they presumed to be the animal's body. The high column descended slowly into the sea as it had risen; and as the last of it submerged the boat began to rock on the commotion of water like the wake of a passing steamer (Maxwell in Heuvelmans 1968:371).

And then there is the other perplexing 1923 encounter with an UMO in Oceania that was observed by three independent groups of people (Heuvelmans 1968:414–415):

> Great was their alarm when they turned round and saw about 60 yards from them a strange animal which gave a prolonged whistle, and shot a jet of vapour, and then a spout of water, vertically into the air to a great height. The animal, whose head stuck out about 30 feet, looked like a sea horse with a crest running well down its back. Its colour was dark mahogany, almost black. The monster made a great wash and undulated on the surface. It rose up and hit the sea with a terrible noise, and then raised itself—head and tail—high above the water again … The animal, which seemed capable of great speed, appeared five times.

> Three times in several minutes I thought I saw several big animals, bigger than porpoises … [A few moments later] I saw several 'bits' bigger than a

sperm-whale, then a black mass: the tail then formed a screen about 6 feet high by 12 feet long. The noise and appearances became more frequent, the black colour remaining… The clearest picture I had of it was of three successive domes, several feet high, which seemed to take up more room than my house. I was a mile away; these three bits seemed to me to be three whales following one another. I cannot describe this monster more exactly, it seemed to me more fish than snake. The length is difficult to judge. The three parts which I saw almost touching measured more than 60 feet and I guessed the animal must extend under water (unless it was a family in Indian file). I did not see the head; but at each appearance, I heard this tremendous noise like the trumpeting of an elephant, followed by the sound of the wash like many sheets of metal falling. It was flat calm.

The animal raised its body vertically like a mast… Sometimes the two branches were held up at the same time like the head and tail of the same animal. These two branches smashed down in opposite directions and in line with one another, making a great noise. Mme Baily confirmed her husband's account in every detail and added that the beast 'frequently spouted a jet of vapour.'

These two examples do read like they could reasonably be concluded to have been observations made of copulating cetaceans that went unrecognized by the lay eyewitnesses. But again, the sizes estimated for the extensions, 7 and 10 m, are certainly longer than whale penises. There is, however, an explanation which might account for the size discrepancy for the extension of the northern Scottish UMO (but not that of the tropical Oceania UMO) and the known dimensions of whale penises. At high latitudes, under conditions of thermal inversion, atmospheric refraction can be substantial enough to magnify and vertically distort distant objects (Rees 1988a, 1988b; Sawatzky and Lehn 1976), which can thereby create aquatic monsters out of otherwise familiar animals (Lehn 1979; Lehn and Schoerder 1981, 2004). But whereas such an illusion might account for optically lengthening a penis to the 7 m observed for the columnar extension of the Scottish UMO, it would be a stretch to believe that this could also be invoked to explain the truly gargantuan "tail" of the Egede UMO. The present theory of the latter being a train of entangled anthropogenic material therefore holds.

CHAPTER 4.

Speculations Concerning the Identity of the Egedes' Entangled UMO

Inferring non-lethal entanglement from accounts of sea serpent/monster encounters is generally an easier task than is postulating the identity of the animal responsible for pulling the debris. This is particularly the case when the encounters are for only a brief duration or when the entangled animal may be a large fish that does not have to repeatedly surface to breathe, as was proposed by de Camp and de Camp (1985) and France (2019a). This chapter examines the possibility that the entangled animal observed off the coast of Greenland in 1734 might have been a basking shark (*Cetorhinnus maximus*).

Identity Candidate—Whale

One of the enduring conundrums of the Egede encounter concerns how much weight should be ascribed to the reverends involved, in terms of their competency as amateur naturalists. Particularly, the question is that if they were so knowledgeable as to be able to distinguish between different species of cetaceans (see below), should we not recognize this and therefore give credence their belief that the UMO they saw was only *like* a whale rather than actually being one per se? There are dozens of sightings of sea serpents/monsters over the years where the eyewitnesses, generally having far less natural history expertise than the Egedes, were able to recognize the presence of large cetaceans (France 2016c). Therefore, the contrasting fact that the Egedes did *not* do so, must certainly count for something. I am aware that this is, of course, the same logic used by cryptozoologists to advance their theories that if the UMO was not a whale, ergo it must be a sea serpent. The difference here is that the animal that I posit might have been behind the UMO is very much a real creature, and one with a long history of being misidentified as a sea monster. In other words, there is no need to immediately invoke mysteriousness to explain mundanity.

At a time when whale populations were much larger than today, there would have been a much greater familiarity with the animals among northern Europeans, both from sightings of living animals as well as from strandings of dead ones (Szabo 2008; Whitaker 1984). For example, 21 folk-categories of whales are described in the thirteenth-century narrative the *King's Mirror*, the first list of cetaceans in world literature (Whitaker 1986). The accuracy of these descriptions are such that the noted polar explorer Nansen (1911:243) commented

> It shows an insight and a faculty of observation which are uncommon, especially in the period; and in many points this remarkable man [i.e., the anonymous author] was centuries before his time. Although well acquainted with much of the earlier mediaeval literature, he has liberated himself to a surprising extent from its fabulous conceptions.

As Szabo (2008:190, 29) notes, "Norwegian fishermen and authors were familiar with many species of whales," so much so that "some individuals within Norse society knew whales well enough to identify species at a glance." Although, hunting pressure resulting in the harvest of thousands of animals would have resulted in diminished cetacean numbers at the time of the Egede sighting, it is likely that this base of taxonomic knowledge would have still been extant. Given their own Arctic natural history interests, the Egedes would likely have been familiar with both the *King's Mirror* as well as the writings of Jon Guomundson, a.k.a. "the Learned," whose seminal 1644 *Natural History of Iceland* (Hermannsson 1924) examined the commonplace folk-cetology (or ethnozoology, if you will) of the residents which had a descriptive accuracy so as to enable him to be able to identify most of the folkloric categories. Indeed, Guomundson's written natural history observations were so remarkably detailed that he was even able to distinguish between resident and transient killer whales (*Orcinus orca*), something that has only recently been recognized by marine biologists (Szabo 2008).

Then, added to all this European knowledge, there is no doubt that through their frequent and close association with the Kalaallit of Greenland, the Egedes would have incorporated that wellspring of local biological knowledge into their own understanding and writing about the region's biodiversity. Inuit today are world-renowned for possessing detailed knowledge about the biology of many species (Ferguson et al. 1998; Gilchrist et al. 2005; Kenrick and Manseau 2008), including marine megafauna (Higdon et al. 2014; Idrobo and Berkes 2012). Although Brown (1868) would disagree, there is little reason to suppose that the same wisdom would have not have existed in the eighteenth-century Inuit as well. So, unlike the situation with regard to the folkloric zeitgeist in which to consider the observed UMO, the Egedes' descriptions of the natural history of recognized Greenland animals, such as various species of cetaceans (see below), were almost certainly influenced by discussions they had with their Inuit neighbors, just as was also the case for their follower, Fabricius, several decades later. Fabricius (1780) documented that of the five species of baleen whales and ten species of toothed whales, four of the former and seven of the latter could be easily identified, and that overall only four species were rarely observed (Kapel 2005). Given the acumen of his former teacher as a capable naturalist, it seems less likely, based on Fabricius' faunistic inventory, that Poul Egede would have failed to identify his "terribly big sea-creature" as a known species of whale should it really have been one.

Paxton et al. (2005:7) ask the obvious question as to whether the UMO seen by the eye-witnesses was "a large unfamiliar baleen whale, perhaps exhibiting a relatively infrequent behavior, for example breaching?" They review the likely candidate species that might have been the animal whose serpentine penis (as they believe) or train of entangled debris (as I suggest) led the reverends into thinking what they saw was some other type of "terribly big sea-creature." In the 1745 translation of his book, Hans Egede certainly starts out correctly by distinguishing between whales and fishes, something, it must be recognized, that was ahead of its time, given that there would be an American court case the following century that would "prove" just the opposite; i.e., that whales were in fact a type of fish (Burnett 2007). In his chapter on sea animals, Egede (through his English translator), confounds interpretations somewhat by referring several times to large cetaceans as "sea monsters." Such phraseology, however, is completely consistent with the medieval and Classical traditions of equating large size with monstrosity (Papadopoulus and Ruscillo 2002; Szabo 2008), and uses the common English vernacular of the age (e.g., Darwin's statement "monstrous as a whale" in *On the Origin of Species*). This is an important point to emphasize. For, from a zoological perspective, to be regarded as a monster means being either: a) an atypically large size (i.e., as used by Darwin); b) an hitherto unknown creature (Westrum 1979) due to it being new to science (i.e., a cryptozoological discovery) or new to the specific location of observation (i.e., a range extension, as discussed below); c) an individual displaying teratological abnormalities or unusual inter-species hybridizations (Dixon and Ruddick 2013; Ritvo 2010); or d) a supernatural or preternatural being (Asma 2009).

In terms of unnaturalness, it is important to recognize that "whales, perhaps more than any other animal, existed in a multiplicity of meanings in medieval thought" (Szabo 2008:23). In this regard, whales were sometimes thought to be malevolent beings under thrall by malignant forces. The upshot of all this is that it was common to conflate whales and sea monsters (Brito et al. 2019; Hendrikx 2018). Given the phylogenetic confusion about what whales actually were biologically, it is not without reason that Vicki Szabo used the phrase "Monstrous Fishes" in the title of her 2008 book about medieval whaling. What this means is that all cryptozoolgical arm-waving ever since about Egede's "most dreadful monster," in addition to the attention-grabbing but somewhat disingenuous usage of that phrase by modern scientists in the titles of their publications (Paxton et al. 2005, and myself herein as well), should be interpreted figuratively given that the English translator was almost certainly confounding all four of the possibilities together with regards to Egede's "monster."

Hans Egede (1741, 1745) provides detailed descriptions of a number of whale types which Paxton et al. (2005) are able to identify: "Finned-whales" are undifferentiated baleen species,

Balaenoptera spp.; "sword-fish" are probably Orcas (*Orcinus orca*); "cachelot or pot-fish" are of course sperm whales (*Physeter catodon*); "white fish" are no doubt belugas (*Deplphinapterus leucas*); "but-heads" are probably northern bottlenose whales (*Hyperoodon ampullatus*); "unicorns" are obviously narwhals (*Monodon monoceros*); and "North Capers" are North Atlantic right whales (*Eubaleana glacialis*). Egede follows this listing with brief mention of porpoises, walruses, and seals, before he moves on to describe the 1734 encounter with the UMO that was something the observers considered to be different from all the aforementioned types of known animals; but not before he first quickly references other sea monsters in Tormoder's *History of Greenland*, which the English translator elaborates for readers to be merfolk, the "hideous" *Hafgufa*, an immense, steam-emitting, legendary creature disguised as an island (it probably was an emerging volcanic islet), the "ghastly" kraken, and the dreaded *Daw*, a supernatural "sea spectre."

Paxton et al. (2005:1) conclude that

> the species seen [i.e., the UMO] was likely to have been [either] a humpback whale (*M. novaeangliae*), a North Atlantic right whale (*Eubaleana glacialis*), or one of the last Atlantic grey whales (*E. robustus*) either without flukes or possibly a male in a state of arousal.

Their conclusions are based on the reasonable idea that the eyewitnesses clearly saw something in the water that was unfamiliar to them. They note that Egede does not specifically describe a humpback whale, unless it is rolled into his catch-all category of "finned whales," thereby making it a possible candidate. However, this species of whale is abundant in Baffin Bay and is remarkably easy to identify given its elongated pectoral fins. It seems unreasonable to imagine that natural historians as accomplished as the Egedes would have failed to recognize such. Paxton et al. suggest that because Egede "does not give any diagnostic characters" and "did not describe it in any detail," the right whale is also a potential candidate for the UMO (Paxton et al. 2005:7). But this ignores the fact that Egede does seem to be quite familiar with this species, given his description of its feeding, distribution, behavior, and anatomy. This species was also regularly found in southern Greenland waters at the time. Paxton et al. (2005:7) are therefore likely correct when they state, "The mention of 'Northcaper' (North Atlantic right whale) by Hans Egede (1741, 1745) may weaken the case of the monster as a right whale because it should have been recognized." And they are on target when they note that Egede did not mention the grey whale, an animal which is now regionally extirpated, and at the time of the UMO sighting "in the 1730s, would have been quite rare and thus may not have been recognized, even if the Egedes were familiar with most species of North Atlantic whale" (Paxton et al. 2005:3). If indeed the eyewitnesses saw one of these last survivors or perhaps an occasional wayward individual from the more healthy Pacific stock which had rounded the southern capes and travelled up to Baffin Bay in 1734, it certainly would have been an anomaly then, just as sighting such an individual off Israel or Namibia is today (Elwen and Gridley 2013; Scheinin et al. 2011).

Given their notably long rostra and pectoral fins, there is no denying the resemblance of breaching grey whales (Figure 4.1) to the body shape of the Egede UMO. But the big difficulty in proposing the UMO to have been a breaching grey whale is that it relies on the supposition that the eyewitnesses not only failed to recognize that particular species of whale (which is certainly possible) but that they did not consider the animal to have been *any* type of cetacean whatsoever. For example, for most of the online photographs which show this species to be breaching, the crease delineating the gigantic mouth of the baleen whale is clearly visible, even from a distance. For some, (e.g., Gould 1930) a belief in the failure of the Egedes to recognize *any* species of whale strains credulity. Was the overt presence of the long tail (whatever it was) such a deal-breaker that the eyewitnesses could simply not countenance that what they saw might have been a whale? Other sightings of cetaceans seen in associa-tion with long serpentine shapes have been interpreted by eyewitnesses as being accounts of said whales locked in mortal combat with sea serpents, as for example, the *Pauline* sighting (France 2016c). Why not then the same for the Egede encounter? After all, it's not as if whales were far from their minds, for both Poul and Hans Egede state in their respective 1741 books that the strange UMO "blew like a whale," and Poul later elaborates in his 1789 book that the UMO's "breath was not strong as the whale's." Herein lies the crux of the mystery concerning the identity of the entangled animal behind the Egede "monster." For clearly Poul Egede, in his original sighting of the creature, and no doubt through later conversation with his natu-

Figure 4.1. Breaching Pacific gray whales. Photos courtesy of Wikimedia Commons. See Photo Credits page.

ralist father, did not consider the UMO to have been a whale. If he had done so, certainly, as cryptozoologists are probably correct in arguing, he would have at the very least mentioned the possibility. Hans Egede too, following on from spending an entire chapter describing the various types of whales seen around Greenland, might have been expected to have done the same. But no, in both cases that which was seen was obviously not considered to have resembled a whale ... of *any* type. On the other hand, there is no denying the compelling bit of adductive reasoning contained in Paxton et al.'s quip that "nonetheless things that blow like whales are, all other things being equal, most likely to be cetaceans" (2005:3). But even here we are still on shaky grounds if we remember that, as Thomas (2011) noted, Bing's original illustration clearly shows the exhalation occurring from the terminus of the snout of the UMO, and not from atop its head, as would be the case for whales. Was this then what finally led the Egedes to consider their "terribly big sea-creature" not to be a cetacean (and Fabricius [1780] to later ignore the possibility)? Hans Egede certainly knew the anatomical dichotomy between the location of cetacean nostrils—earlier in his book, for example, he describes the breathing holes of narwhal to be situated atop their heads—and those he described for the enigmatic UMO.

Despite all these arguments, there is a reasonable likelihood that Paxton et al. (2005) come closest to hitting the mark when they suppose that the observed animal might have been a seldom seen grey whale. One only has to look again at Figure 4.1 and other online photographs of such breaching individuals to feel comfortable in endorsing such an illation. However, for the rest of this chapter, I will play devil's advocate and posit an alternative hypothesis as to "the nature of the beast" with respect to Egede's mysterious creature.

Identity Candidate—Basking Shark

As an alternative to the cetacean hypothesis, is the possibility that the Egede UMO is another example of the basking shark's alter ego as natural history's most widely misconstrued sea monster. Following on from his discussion of whales as "monsters" and the mysterious "terribly big sea creature," Hans Egede (1741, 1745) continues his faunistic survey of Greenland by describing a variety of fishes such as halibut, haddock, salmon, etc., including the biggest of all, the *Hak*, or Greenland shark (*Somniosus microcephalus*). And just as for the Atlantic grey whale, no mention is made of the basking shark (*C. maximus*), another visitor to southern Greenland waters. This species can grow in excess of 10 m (the size of a school bus) and is the second largest fish in existence today (Fairfax 1998; Speedie 2017). Anatomically, basking sharks possess large pectoral fins (one-and-a-half meters in length), two dorsal fins (each fully a meter in height), a pair of meter-long claspers on the inner borders of male pelvic fins which are used during copulation, a pointed rostrum, and an oil-filled liver which can account for a quarter of the overall body weight and which is used for hydrostatic buoyancy (and also prized as a lubricant). As well, they are noted for having a deeply fissured skin that is covered with sharp denticles, referred to as placoid scales. Being filter-feeding plank-tivores, basking sharks are often observed swimming close to shore, especially headlands, with mouths agape more than a meter wide. They swim at slow speeds of 4–7 km/hour near the surface while zig-zagging their way through dense patches of plankton. Although often giving the impression that they are almost stationary and basking in the sun, the sharks can sustain non-feeding speeds in excess of 9 km/hour.

Basking sharks were late to enter the natural history record. Their first mention, identified as *bein-hakall*, occurs in Jon Olafsson's *Icthyographica Islandica*, an early eighteenth-century treatise on the fish fauna of Iceland (Fairfax 1998). Pontoppidan (1755:109, 116) refers to "*haae*, the shark," and *haae-maeren* or *brigde*, as being notable for their oil, but confusingly aligns them as being akin to cetaceans rather than elasmobranchs. It was not until Johan Ernst Gunnerus, the Bishop of Trondheim and an experienced amateur natural historian, wrote in 1765, at a time coincident with a rapidly expanded fishery of the sharks, that the species, referred to by him as *brugden*, was formally described in the scientific literature. He distinguished it from the Greenland shark, determining it to be a new species, *Squalus maximus*, that fed on "tiny insects" in the water, as plankton were then quaintly called. This, it is important to note, is decades *after* the Egedes referred to their mysterious UMO as a "very horrible sea-creature." Even today, the elusive species remains largely enigmatic; for example: "little is known, however, about many aspects of the life history and biology of bask-ing sharks" (Cotton et al. 2005:151).

No species has had a closer ethnozoolgical relationship to sea monsters than the bask-ing shark. In his chronology of Nova Scotian UMO sightings, Hebda (2015:35) states: "It is of interest that the Basking Shark plays an on-going role in this sea monster saga, perhaps due to its size, unusual anatomy or its rarity in encounters with the casual observer." Such a conjoining of fact and fantasy has long been recognized. With reference to Magnus' *Carta marina*, for example, "it is believed that the basking shark inspired creation of some of these wonderful creatures" (Lilja Bye 2018). And one must not forget that the final accepted Latin binomial name for the species, *Cetorhinus maximus*, denotes "pointy-nosed monster of great size"; i.e., the first part of the genus name *Ceto* is derived from the Greek *ketos*, meaning marine monster (Papadopoulis and Ruscillo 2002; see Appendix 1.4). It is not without due cause, then, that Colin Speedie (2017) entitled his book *A Sea Monster's Tale: In Search of the Basking Shark.*

Today, despite change in the perception of basking sharks from monsters to charis-matic megafauna (Mazzoldi et al. 2019), they continue to receive bad PR through being closely associated with the former. The species is mentioned directly in the titles of news articles or prominently referred to in the texts of others. A lecture was delivered in Dublin in January 2020 and an earlier article about the species' return to British waters (Renton 2013) continued to remind that "behind most of the Atlantic coast's myths of water monsters and sea snakes lie basking sharks with their weird snouts and confusing skeletal remains."

Often peaks in sea serpent sightings, such as those that occurred in the middle of the twentieth century in British Columbia, Canada, correspond to periods of high abundance of basking sharks (Speedie 2017; Wallace and Gisborne 2006). Other observers (e.g., Magin 1996) have noted that the annual spring migration of basking sharks along the British coast corresponds to the first seasonal reports of sea serpents, and after the sharks depart the area, the sightings cease. Part of this is no doubt due to the propensity of the animals to swim in long lines, nose to tail, such that "when seen from a distance they look like nothing other than a sea snake or plesiosaur" (Speedie 2017:18). Speedie provides a modern anecdote of standing on the shore beside an individual whom insisted the distance view of a basking shark off the Scottish coast was that of the sea serpent she had often observed thereabouts.

Numerous reports of UMOs from the British Isles (Harrison 2001) also suggest unrecognized basking sharks. Some seem to describe normal plankton feeding: the creature possessing a huge great mouth, described as being white on the inside; or a head of long, tapering shape with an enormous, open mouth filled with shining teeth; or a long low dark object skimming along the surface of the sea while displaying its back. Other reports describe normal breaching behavior characteristic for the species (see below): the UMO causing a great disturbance before finally leaping straight out of the water straight as an arrow; or another monster that suddenly shot out of the water into the air to a height of about 15 m; and another that lashed the surface with its body before it raised its fore part out of the water for fully 4 m, followed by flopping back down with a loud crash, an act that was repeated several times.

Mostly, it is the unusual and prodigious size of the basking shark that attracts attention and makes observers think and refer to them as being "monstrous" (Figure 4.2). Gavin Maxwell, for example, himself a witness to one of the most famous sightings of a purported sea monster (France 2017), describes one of his first encounters with a basking shark, whose shape he likens to the body of a dragon. In particular, he was struck by the size of the two pectoral fins which were much larger than what he had seen in various illustrations. As Maxwell recounts (in Speedie 2017:96), the naturalist was impressed by the sheer magnitude of the harpooned animal as it was raised tail-first out of the water with a crane:

> Size always appears greater in the vertical then the horizontal [this is something to keep in mind with respect to the vertical height estimated for the Egede UMO's breach], and by the time fifteen feet of the shark were clear of the water and the girth was still increasing, he appeared literally monstrous, a creature of saga or fantasy; a dragon being hauled from its lair. The darkness, the shifting yellow reflections of the harbor lights, and the white glare

Figure 4.2. A freshly caught basking shark is transported to dry land in waters off the Western Hebrides, Scotland. Original Publication: Picture Post - 4209 - Shark Hunting - pub. 21st September 1946. Photo by Raymond Kleboe/Picture Post/Hulton Archive/Getty Images.

of the searchlights combined to give a stage effect of mystery and magnification. There was an excited gabble from the packed crowds on the pier, gasps and exclamations, and a group of women near the edge panicked and forced their way back into the press behind them.

'Oh, what a crayture!'

'Ye woulna' believe it!'

'It canna be a fish!'

Maxwell (1952) believed that many so-called sea monster sightings were really enormous basking sharks, noting that when the animals feed near the surface, the tips of their snout, anterior dorsal fin, and upper tail lobe are all exposed conspicuously out of the water, which, given the size of the animal, are often separated by a distance (Figure 4.3). For example, Heuvelmans (1968:369) offers the following matter-of-fact synopsis of a 1903 sighting in Scotland: "They said that a little head and two big fins appeared out of the water—which is just what one sees when a basking shark basks." And as another example, Maxwell focuses on the H.M.S. *Hillary* encounter with a 20 m long UMO off the coast of Iceland in 1917. Alerted to the fact that an object had been seen in the water that was thought to be living but, as for the Egedes, not to be a whale, the captain sees something which

at first glance suggested to my mind a tree trunk with only the knobby ends (from which branches and roots had been cut) visible. A careful look through my glasses, however, made it clear that the thing was alive, and the 'knobby ends' were in fact its head and dorsal fin (Heuvelmens 1968:396; Figure 4.4A).

Figure 4.3. Demonstration of how the above-surface exposed tips of snout, anterior dorsal fin, and upper tail lobe of a surface-feeding basking shark can give the impression of an elongated mystery animal of large size (Maxwell in Fairfax 1998:184).

Moving the ship closer, they determined that the head was about the size of that of a cow, did not display any horns or ears, had a white strip between the nostrils, and bobbed up and down. As well,

> from the back of the head to the dorsal fin no part of the creature showed
> above the water, but the top edge of the neck was just level with the sur-
> face, and its snake-like movements could be clearly seen [Gould (1930)
> interprets this to mean that the UMO curved itself into a semi-circle]. The
> dorsal fin was a black triangle that rose more than a meter above the water,
> and the creature did not seem to be in the least alarmed by the presence of

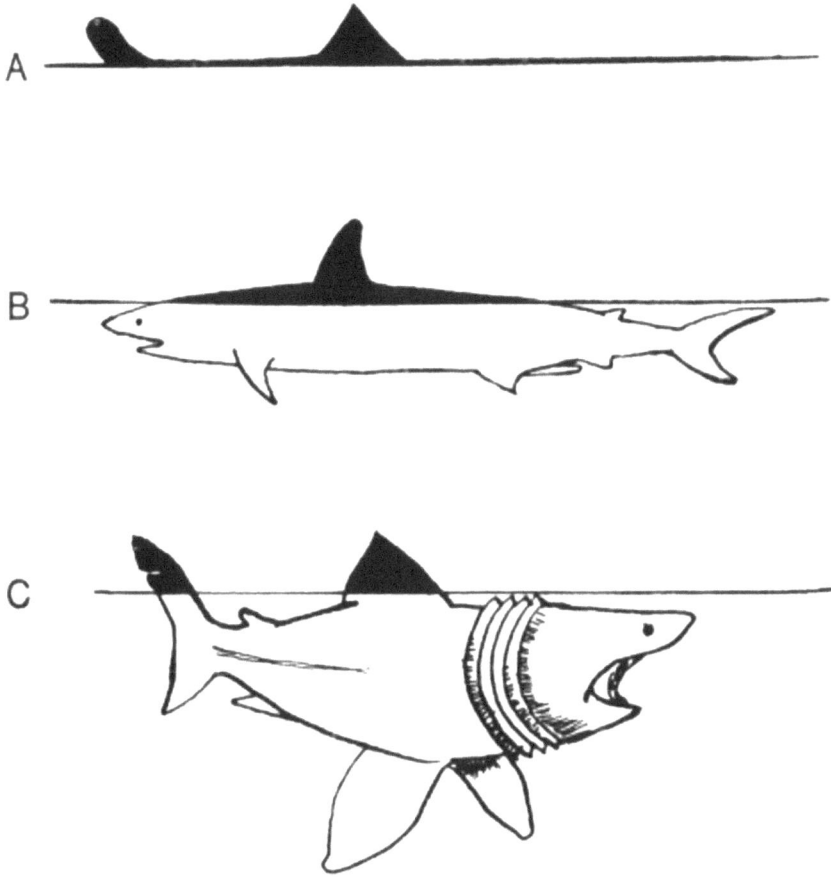

Figure 4.4. The 1917 sighting by the H.M.S. *Hillary* off the coast of Iceland of an UMO (upper panel), re-interpreted as a basking shark by both Gould (1930; middle panel) and Gavin Maxwell (lower panel; in Heuvelmans 1968:399).

the ship but continued to bask on the surface, now sinking down till only the tip of its nose and fin were visible, and anon rising again (Heuvelmens 1968:396–397).

Gould wrote to the captain and asked him whether he might have mistaken a basking shark for the observed creature, including a drawing of such (Figure 4.4B) and mentioning that he personally has seen these animals swimming flush with the surface in slow, graceful curves with only the dorsal fin showing. The captain responded, emphasizing that the creature was "*most certainly not a shark*" (Gould 1930:213, emphasis in original), leaving Gould, who notes the similarity of its swimming behavior to that of the famous Gloucester Sea Serpent, to conclude that the creature was something unknown to science. However, Maxwell (1952), who, unlike Gould, was a competent naturalist (he was the author of the classic otter tale, *Ring of Bright Water*), will have none of it. He considers Gould's drawing to be inaccurate and offers his own explanation of how a basking shark can resemble the UMO (Figure 4.4C), adding that "nor in his [the captain's] lengthy description can I see anything but my old friend the Basking Shark" (Maxwell 1952:214). Furthermore, as noted by Maxwell, and later by Ellis (1994), many basking sharks actually do have a white strip on the underside of their snouts. Like many a cryptozoologist reticent to accept mundanity over mysteriousness, Heuvelmans (1968:400) disagreed, confident in his belief that "the Hilary monster, in fact, seems like our old friend the New England sea-serpent, caught once again basking on the surface." By this he means the 1817 Gloucester UMO, whose train of entangled fishing floats, misidentified as the elongated body or tail of the creature, was often observed to be lying or basking on the surface while the animal (probably a large fish like a giant tuna) was submerged for long periods (France 2019a, 2019b).

In 1808, a carcass was found on the Orkney Island of Stonsa (now Stronsay) that was 18 m long and a meter thick, and which had a small head, a horse-like mane of bristles, and three sets of odd appendages shaped like paws, which would later be described by some as "wings." At a meeting of the Wernerian Natural History Society, an offshoot of the Royal Society of Edinburgh, and as reported in the Society's *Philosophical Magazine* of November 19, 1808, Secretary Patric Neill read a letter about what would soon become known the world-over as "The Stonsa Beast." As Oudemans (2007[1892]:60) wrote,

> [Neill] concluded with remarking that no doubt could be entertained that this was the kind of animal described by Ramus, Egede, and Pontoppidan, but which scientific and systematic naturalists had hitherto rejected as spurious and ideal.

The newly encountered name here, "Ramus," refers to Johann Ramus, who described a 1687 Norwegian sighting sea serpent which I believe to be one of the first clearly described instances of an entangled animal (see France 2019a). The original quotation from Neill was about finding a "singular Animal, of great size, and corresponding to the description given by Egede and Pontoppidan, of the Great Sea Snake of the Northern Ocean" (Fairfax 1998:45–46). Preceding the legalistic-styled investigation of the Gloucester Sea Serpent (Soini 2010) by a decade, the Society collected affidavits of eyewitnesses and solicited a drawing of the remains of the creature (Figure 4.5). The naturalists labelled the Stronsa Beast as a new species: *Halsydrus pontoppidan* (i.e., saltwater snake of bishop Pontoppidan). And so a monster was born (Swinney 1983) that "entered the world of cryptozoology" (Speedie 2017:17). Newspapers and magazines around the world were beside themselves in publicizing the first definitive physical proof for the existence of a sea monster.

The beast's attributes as described by eyewitnesses were published in the premier issue of the Society's *Memoirs* (Barclay 1811), one trait being the presence of soft teeth which were malleable by hand. But this was not before amateur natural historian Sir Everard Home, M.D.,

Figure 4.5. Drawings of the 1808 Stonsa Beast "globster" by an eyewitness (top panel photo courtesy of Wikimedia Commons; see Photo Credits page), a commissioned artist (middle panel; Home 1809:212), and as rendered by Oudemans (2007[1892]:159; lower panel), and which turned out to be a decomposing carcass of a basking shark rather than a sea monster.

who was the first to describe the ichthyosaur fossils found by Mary Anning and her brother (see Appendix 1.1), examined the remains and recognized them to be those of a basking shark. The Society had sent the affidavits to Sir Joseph Banks, who in turn, had passed them onto Home. Coincidently, Home had been working on an anatomical study of the basking shark at the time, through examining the body of one which, interestingly for the purpose of my present thesis, had become "entangled in herring nets belonging to fishermen" (Heuvelmans 1968:123). The misidentification of the remains, Home (1809) explained, was due to differential decomposition, whereby the soft cartilage of the massive gill-rakers and portions of the fins disappeared first, leaving behind a mass (today such remains are referred to by the wonderfully onomatopoeian word "globster") of little else than the small cranium attached to the spinal column. The result is a carcass which does not in the least resemble a shark but rather does that of a plesiosaur with a long and skinny neck. Furthermore, Home noted that the spinal contortions depicted by the illustrator of the carcass were anatomically impossible, and that it was "highly probable that the account of Pontoppidan's sea snake had been read by the spectators of this fish, in the interval of time between their seeing it and their depositions being taken" (Home 1809:218). In short, his examination of the vertebra left no doubt that the Stronsa Beast was indeed a basking shark.

Moreover, the third set of dog-like hind "legs" of the Stronsa Beast (unlike anything in the known zoological world) was due to the presence of the male claspers or holders of the basking shark used during copulation (Figure 4.6). In this regard, the "veritable wonder" of

Figure 4.6. Victorian onlookers marvel at a stranded basking shark. Note how the male's copulatory claspers have been interpreted as a set of legs and feet (Harper's Weekly, 24 December 1968:24).

basking shark globsters, both the Stronsa Beast and the many others that would follow, contributed much to fostering an abiding belief in oceans populated by mysterious megafauna (Rotschafer 2014). By infusing meaning into natural history, printed illustrations such as the basking shark in Figure 4.6—which Rotschalfer uses to establish her thesis—played a critical role in the nineteenth-century sea serpent craze (see Chapter 5).

The Society's John Barclay, M.D., himself a professor of comparative anatomy, cried foul, expressing his belief in the 1811 article that no shark could have such a long neck. As described by Oudemans (2007[1892]:68):

> Mr. Home concludes by observing, that 'it is of importance to science, that it should be ascertained, that this fish is not a new animal, unlike any of the ordinary productions of nature.' Of what importance it is to science to admit no new genera or species into our catalogues of natural history, I cannot conceive. But it is certainly of much importance to science, that the naturalist should be cautious not to determine the species of an animal upon vague evidence. Now what evidence had Mr. Home that this animal was a squalus [i.e., a shark], and even to suppose that it was a squalus maximus [i.e., a basking shark; the species' binomial name at the time]?

The irony here concerning the charge for naturalists to express caution before jumping to conclusions is precious indeed; not just with respect to the Stonsa Beast, but to the larger field of cryptozoology itself (i.e., given Oudemans' significant role in fostering the belief in sea monsters).

Thus the debate began, which would continue for much of the nineteenth century, in the pages of journals such as *Isis*, *The Proceedings of the Royal Society of Edinburgh*, *The Zoologist*, *Archives fur Naturgeschichte*, and books. Leading naturalists of the day argued back and forth as to whether the UMO's remains were those of a basking shark or something else, such as perhaps a plesiosaur, whose fossilized remains were just then becoming widely known. Scottish naturalists, unable to objectively free themselves of nationalistic pride, continued to refer to the "Orkney Sea Snake" well into the middle of the century, at the time of the debate about the *Daedalus* UMO sighting (Traill 1854).

Even as late as the 1930s, zoologists were still going back and forth on the identity of the Stonsa Beast; this despite other globsters having washed ashore in the many years since, all of which shared in common the same time course: initial speculation in the press that finally a sea monster has been discovered; then, after only a few days and subsequent physical examination by a trained zoologist, the dreams of crypotozoologists dashed with announcement that the remains were, yet again, those of a basking shark (e.g., Hebda 2015; Kuban 1997). Even the publication of illustrations that repeatedly showed how the latest plesiosaur-like sea monster discovery, as for example by Petit in 1934 concerning the "Cherbourg monster" (reprinted in Heuvelmans 1968) or the 1977 sea monster hauled aboard the Japanese fishing vessel *Zuiyo Maru* (see Loxton and Prothero 2015), could be created by the differential decomposition of tissues from basking shark carcasses, failed to dissuade the "truth-is-out-there" believers. During the time in 1933 when "Nessie" was being invented (Jylkka 2018; Williams 2015), following the premier of *King Kong* (Loxton and Prothero 2015), Scottish professor of natural history James Ritchie reexamined a vertebrae of the Stronsa Beast that had been stored at the Royal Scottish Museum, and concluded that:

> the sea-serpent of Stonsay, which a century and a quarter ago raised so great a commotion in the scientific world, has fallen from its unique estate, but it remains a not-to-be-forgotten memorial to the credulity of the inexperienced and of the scientists who built upon so shaky a foundations (in Loxton and Prothero 2015:215).

To which Loxton and Prothero (2015:215) remark, "The Stronsay Beast may never have been quite forgotten (at least within the niche sea serpent literature), but its lesson has never been learned. Again and again, people have fallen for the same grisly illusion," resulting in one of "cryptozoology's silliest banalities." As an absurd example, they describe how so-called "scientific creationists" (i.e., Old Testament literalists) have argued that the *Zuiyo Maru* carcass was not that of a basking shark but really was a plesiosaur. Doing so thereby allowed creationists to prove the fossil record to be problematic and evolutionary theory itself to be suspect. As Fairfax (1998:49) aptly concludes:

> These 'sea-monsters' cast ashore from time to time will always excite comment and the mystery lingers in popular perception long after the clear light of scientific analysis has established the true nature of the creature. It would seem that we need our monsters and are reluctant to see them debunked.

My belief is that a number of the sightings of sea monsters, including that of the famous *Leda* UMO off the northwest coast of Scotland in 1873 (France 2019c), which is "the most detailed account of a sea serpent encounter" in British waters (Harrison 2001:153), can be parsimoniously explained by basking sharks pulling trains of entangled fishing net floats or harpoon line kegs. During the mid-1990s, while taking the ferry between Vancouver and Victoria, British Columbia, Canada, I saw what first looked like a classic many-humped sea serpent, my expectant attentive imagination momentarily kicking in as this was the reputed home territory of the famous "Cadborosaurus" sea creature (LeBlond and Bousfield 1995). However, closer inspection as the ship moved past revealed the UMO to be what looked like a large, slow-moving shark that was pulling a string of fishing net floats bobbing up and down along the surface of the water. If this was indeed one of the—at the time and still today—rare basking sharks in the region, seeing such an entangled animal would not have been unusual given the species' noted proclivity for becoming entangled in fishing gear the world over. Indeed, during the 1940s in British Columbia, the problem became so severe, with hundreds of salmon nets being ruined each month, that the government declared basking sharks to be "destructive pests," and instigated a fishery to cull them, which lasted until 1970 (Wallace and Gisborne 2006). MacNeil et al. (2012) believe that some of the bycatch recorded for Greenland sharks, particularly that which occurs in north-temperate climes, may actually be misidentified basking sharks. The Irish actually capitalized on this predilection for basking sharks to become entangled when they established their own specialized net-based fishery for the animals. In the recent past, the statistics have been gruesome, the accidental bycatch of basking sharks in gill-nets exceeding hundreds per annum (Lien and Fawcett 1986), leading to many attempts to free such animals. Thomas (2011) recounts several intriguing anecdotes obtained from the summer he spent working at the University of Copenhagen's research station on the island of Disko, in Disko Bay, West Greenland, located near where the original Egede encounter occurred. Noting that "the Greenland sleeper shark [*S. microcephalus;* which can grow to lengths of 7 m] may be the reason for at least some of the monsters sightings in Greenland waters" (Thomas 2011:14), he goes on to describe how globster remains that had washed ashore in the 1990s were analyzed and found to be those of a shark, and also how in 1940 one woman recovered the remains of "what looked like a hairy whale" that had become entangled in a fishing net. "Hair" in this case might very well be either the seaweed-festooned netting material itself or perhaps the long strands of cartilage that are often displayed on decomposing filter-feeding sharks. Since the late 1980s, a long-line fishery has existed for Greenland halibut which has resulted in thousands of Greenland sharks falling victim to becoming bycatch (Idrobo and Berkes 2012), including off Disko Island itself (MacNeil et al. 2012).

The entanglement of basking sharks has long been noted. As far back as the eighteenth century, Pontoppidan (1755:109) makes reference to how Norwegian fishermen were "as much afraid of them [i.e., *brigde*] as of the most dangerous sea-monster" when the sharks circled their boats and became caught in their nets. This is one of the earliest descriptions of bycatch. Fairfax (1998:61) reproduces text from an 1801 London advertisement, entitled "A Nondescript Or, WONDER OF THE DEEP," wherein the curious public, for the price of one Shilling, could observe the "extraordinary" giant shark on display, "after it was entangled in the Net... [and then] received SEVENTEEN MUSKET BALLS, and many other Wounds before it expired." This is one of the earliest uses of the word "entangled" with respect to fishing nets. Similarly, a report in a Scottish periodical from a decade before gives one of the first mentions of the word "disentangling" in the literature, in this case with regard to basking shark bycatch:

> This sluggish fish sometimes swims into the salmon nets, and suffers itself to be drawn towards the shore, without any resistance, till it gets near the land, that for want of a sufficient body of water, it cannot exert its strength, in disentangling itself from the net, the fishers in the mean time take advantage of its situation, and attack it with sticks and stones, till they have it secure (Fairfax 1998:84).

And Heuvelmans (1968:369) summarizes another encounter that occurred in Scotland in 1903 that resulted in ten nets being "destroyed" by an UMO: "All the details, the nose ending in a point like a horn, the triangular dorsal fin, the size, the indolent nature, the mucus-covered skin, the torn nets—agree perfectly with its being a basking shark."

Of particular note is the case of the "sea serpent" observed in 1825 in the inner harbor of Halifax, Nova Scotia. Eyewitness descriptions of the UMO (Hebda 2015:31–32) strongly suggest an entangled animal that,

> impel[led] itself forward without fins and by a wriggling motion of the body... raised a coil which warped itself along to the tail, and this made it to be supposed that its extreme length was at least 60 ft... [such that] at one time eight coils of the body [were seen] above the water, each about a

yard in length and with the same space intervening between … [when] it moved through the water with great rapidity, and left a large troubled wake. It showed no fins.

The UMO became much discussed, some opining it to be a line of porpoises, and others, a black log. However, the mystery was cleared up a few days later, when the (or a similar) UMO was caught at the head of the inlet, 20 km away:

> The report of the appearance of the Serpent gained very strong confirmation from news having reached the town on Friday morning last, that a sea animal of unknown species had been caught at Portuguese Cove – but upon inspecting it, it turned out to be of the species *Glaucus maximus or Basking Shark* … It was caught by having entangled itself in fishing nets, which were set for mackerel … The persons who witnessed it say, that its struggles to free itself were tremendous, often rearing itself twenty feet from the surface of the water, and lashing all around it into froth and fury. We were informed by the inhabitants on the spot, that there had been caught before at least five or six of the same species (Hebda 2015:33, 35; emphasis in original).

The description of this unfortunate entangled shark in throwing itself into the air and thrashing about is most significant.

The early basking shark fishery to collect the oil contained in their livers was opportunistic. But by the end of the twentieth century, this had grown to more than a hundred thousand animals being killed. Today, basking sharks are red-listed as a threatened species by the ICUN (International Union for Conservation of Nature) and, at least in Europe and North America, are pursued for ecotourism purposes rather than extractive harvesting (Mazzoldi et al. 2019). Once harpooned, struck animals would often display considerable strength and tenacity. Fairfax (1998) and Speedie (2017) describe incidents with harpooned sharks pulling, with "amazing rapidity" and "great violence," both a dingy and even a moderately-sized vessel of 70 tons (63,500 kg). Such "Nantucket sleigh rides" of pulled dories, to use the parlance of Massachusetts whalers, could often last for more than half a day. And, as was also the case for tuna hunting, basking shark fishermen sometimes attached a string of buoys, barrels, or drums to the harpoon line in an attempt to create enough resistance to slow down struck animals.

The 1825 Nova Scotian report (Hebda 2015) of the entangled basking shark's energetic attempts to free itself from the mackerel net noted above seems to have been forgotten. This has led to the common perception of the animal as being a slow moving, dim witted, gentle giant. Such a misconception was challenged, however, by what became known as "The Carradale Incident" of 1937, which occurred in western Scotland. At a time when numerous basking sharks were observed "leaping out of the water non-stop" (Fairfax 1998:83), one repeatedly breaching animal capsized a small boat, which resulted in the drowning of its three occupants. This incident, too, seems not to have made an impression on biologists. Despite a British zoologist mentioning breaching of basking sharks as far back as 1769 (Fairfax 1998), the behavior is rare enough that, even as late as the middle of the twentieth century, biologists still questioned whether such behavior was characteristic for the species. Matthews and Parker (1950), for example, believed that reports of basking sharks leaping so high as to completely clear their bodies of the water were incorrect. We now know that it is this conclusion itself that is incorrect (Gore et al. 2018). Recent work involving analysis of filmed breaches and individual-borne data loggers has shown that these famously languid planktivores can reach vertical velocities of 5 m/second when ascending from a depth of 28 m, allowing them to break the surface of the water at steep angles in only nine seconds and ten tail beats (Johnston et al. 2018). Speedie (2017:41) describes a personal encounter in which a basking shark with no spatial awareness of his boat was seen "erupting like a missile out of nowhere, so close to our yacht that the splash on re-entry sprayed water all over." Another incident was observed in Scottish waters by Greenland explorer Anthony Watkins:

> Suddenly there was a mighty rushing sound and a huge glistening shape rose out of the water at a distance of less than forty yards, stood poised on its tail, high as a three storied house, for a fraction of a second, then fell forward in a thunderous belly-flop to disappear in a great mountain of foam (Speedie 2017:48).

This reads remarkably similar to the 1734 Egede encounter. And there is no doubt that online images of breaching basking sharks most certainly do resemble the Egede UMO (Figure 4.7). Here can be seen the fusiform bullet-like shape of the animals, as well as their prominently displayed huge pectoral fins. Also, and unlike the case for breaching grey whales, the aerially-borne basking sharks frequently exhibit their large, anterior dorsal fins. Furthermore, given the great speed with which they erupt out of the water, they rise higher and display much more of their bodies than do whales. The result is that they sometimes also show their posterior dorsal fin as well as even their pelvic fins. Given this, it is easy to speculate that one of these fins might be indicative of the presence of the protuberance shown in Bing's illustration. The Egedes can certainly be forgiven for being fooled if what they saw was indeed breaching basking shark, for as Bright (1989:180) states, it is the type of animal that is the most likely to be mistaken for something more unusual, such as a sea monster, "particularly when it leaps clear of the water."

Some breachings of basking sharks involve solitary sharks, whereas others occur in association with other animals. Multiple breachings in quick succession are known (e.g., the

Figure 4.7. Breaching basking sharks. Top and bottom-right photos courtesy of Anthony Robson. Bottom-left photo courtesy of Bren Whelan. Used with permission. See Photo Credits page.

infamous Carridale Incident) and are often taken to be two fish, although Fairfax (1998:31) notes that "in the flurry of spray it would be very difficult to distinguish whether there was only one fish present." Remember that Poul Egede described the UMO as breaching three times. Breaching's ultimate purpose is unknown but might possibly be a form of social communication related to courting (Gore et al. 2019) or a process of dislodging ectoparasites such as lampreys or copepods. What is known is that a breaching basking shark does *not* resemble a breaching whale. Cetacean breaching, regardless of species, Speedie states—having observed hundreds of such incidents—is characterized by acrobatic grace, power, and a "balletic élan." Not so, the basking shark, which "simply launches itself through the surface of the sea like an unstable missile, twisting as it goes…before crashing down on its back with a massive bang and a cascade of spray" (Speedie 2017:200). Additionally, it is interesting to note that one accompanying behavioral trait displayed by breaching basking sharks is the ferocious lashing of their tails following the aerial maneuver. Again, this might be due to furthering the process of communication or parasite dislodgment. Regardless of the ultimate motive for breaching, should a string of entangled debris be attached to said tail thrashing about, it is easy to imagine that there would be instances when it would be waved about in the air, just as Bing's drawings indicate, and as other UMO sightings of what I believe to have been entangled animals show (Figures 3.1 and 3.3). Remember, too, the aforementioned account from 1825 of a basking shark that launched itself fully "twenty feet from the surface of the water, and lashing all around it into froth and fury" (Hebda 2015:35) in its vain attempt to free itself from entanglement in a Nova Scotian mackerel net, and one can see how a so-entangled shark might be mistaken for an unknown sea monster. Finally, unlike breaching whales, basking sharks doing the same show absolutely no spatial awareness in general, or avoidance of boats in particular. Again, this is similar to the noted proximity of the ship to the breaching UMO as described in the Egede encounter.

At first blush, a major drawback to the illation that the animal behind the Egede UMO may have been a basking shark, concerns the vaporous exhalation from its head, which, Polonius-like, Egede stated as being "[very] like a whale." Basking sharks of course do not surface and exhale before taking in fresh air. However, perhaps criticism has been too harsh with regards to Pontoppidan's illustration, wherein the cloud of vapor shown in Bing's original drawing (Figure 1.7) was transformed into a stream of water (Figure 1.11). Burnett (2007:126) quotes a nineteenth-century naturalist who spent time with whalers and become amazed at their ability to distinguish different species based on the "shape of the vapor from their exhalations, which from a certain distance resemble (indeed, could be taken for) jets of water." Certainly, when one looks at drawings from the period used as an identification key to whale exhalation (Browne 1846; Figure 4.8), these more closely resemble Pontoppidan's geyser of water than they do Bing's exhalation of vapor. And of course, the same pertains to all those early depictions of monstrous whales on Renaissance maps (e.g., Figures 1.1 and 1.2). Furthermore, Fairfax (1998:84) recounts an eighteenth-century referral to basking sharks "rising and blowing" near headlands. So is it possible that the vapor exhalation might have actually been nothing more than a spray of water associated with the Egede UMO's breaching? Online photographs of both breaching grey whales and basking sharks (e.g., Figures 4.1 and

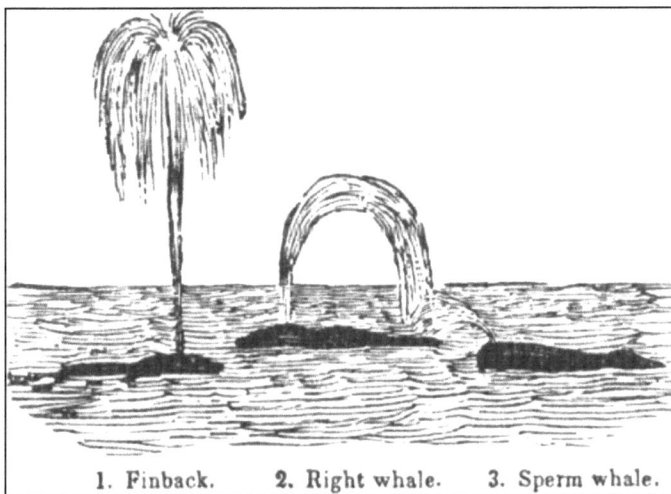

Figure 4.8. Species identification of whales based on the typology of their "blow" (Browne 1846:281), a procedure which Burnett (2007) refers to as "reading the mist."

4.7) show water spray arising from the heads, sometimes near the tip of the snouts, for both species. This might explain the incongruity, as noted by Thomas (2011), of Bing's illustration, which shows vapor, or spray, emerging from the terminus of the snout and not from the top of the head of the UMO (something, it should be noted, that even the early Renaissance map-illustrators got right in their exaggerated renderings of monstrous whales). It should be noted that cetaceans generally do not exhale during the act of breaching or spy-hopping. This could suggest that it was more likely to have been a spray of water and not vaporous exhalation that was thrown into the air by the breaching Egede UMO, regardless of whether the animal was a basking shark or an unrecognized species of whale.

A personal anecdote may offer some support to the present contention. In 1990, at the termination of the first north-to-south, non-Indigenous ski-crossing of Ellesmere Island (France 2010), and about 2000 km from the southern tip of Greenland, I observed a breaching, bullet-shaped animal at too great a distance to make a positive identification. What I was convinced of at the time was that it did not resemble any film footage I had ever seen of a breaching cetacean. I remained watching the seascape of the great polynya of Baffin Bay (France and Sharp 1995) for more than an hour, but the mysterious animal never reappeared, thereby tentatively suggesting that it was not a whale. Speedie (2017:230) describes a similar encounter in Scotland: "Often the breaching animal would be some distance away, and so positive identification as a Basking Shark was not always possible, but the chaotic style of the breach and the lack of any further sighting almost certainly ruled out any cetacean species"; as does Fairfax (1998:31): "A breaching basking shark seen at a distance could be confused with a small whale such as a minke but the absence of spouting before or after the event strongly suggests that the animal in question is a basking shark." Wildlife observations during the ski expedition had already extended the known distributional range of one species by thousands of kilometers (France and Sharp 1992). Could this sighting of a breaching UMO be another northward range extension for the basking shark as well?

Basking sharks normally migrate thousands of kilometers, and are capable of covering prodigious distances, such as crossing the Atlantic Ocean in the span of three months (Speedie 2017). Some basking sharks are now known to undertake extended journeys. One was a victim of bycatch in 2015 in Australia, the first presence there in three-quarters of a century (Australian 2019). The fact is that we have only an incomplete knowledge of the true distribution range of basking sharks, thereby making any conclusions about residency or waywardness problematic. For example, of two dozen basking sharks tagged off New England, eight stayed within the previously described range of the species based on earlier studies, but the rest were found to expand the known range (Skomal et al. 2009). Speedie mentions that there has been a gradual northward shift of basking sharks, with more sightings than ever before occurring in Iceland. Other work has shown the species to have recently become abundant in northern Norway (Lilja Bye 2018). Might the mysterious animals seen in Baffin Bay by both Poul Egede and Bing in 1734, and by myself in 1990, have both been wayward basking sharks? Given that the regional abundance of basking sharks is correlated with surface sea temperatures (Cotton et al. 2005), it might be easier to accept such an illation for the twentieth-century sighting than for the eighteenth-century one, given that mine took place during a time of climatic warming whereas Egede's occurred right at the peak of the Little Ice Age.

Sea monsters are born of seeing things that are anomalous or preternatural. This can be, as argued above, created by maritime debris entangled around what would otherwise have been a recognizable animal, the resulting hybrid looking like something that is strange and different. Strangeness can also occur when a perfectly normal-looking animal is simply seen outside of its typical distributional range. The unfamiliarity of the creature being observed can often lead to the unfounded conclusion that it must be something completely unusual to natural history, whereas in fact it is only its temporary presence in that particular area that is unusual. For example, putative sea monsters in both Chesapeake Bay and Massachusetts have been hypothesized to have been vagrants. This is given credence by the fact that in recent years a manatee and a beluga whale have both been observed in New England, the former swimming all the way up the coast from Florida, the latter taking a long journey south from Quebec or Labrador.

Basking sharks are widely distributed in the temperate boreal waters of the North Atlantic (Skomal et al. 2009). They make use of thermal fronts as foraging and migration corridors (Sims et al. 2003), and are customarily found in a wide arc that tracks the Gulf Stream in stretching up from New England, through Atlantic Canada, brushing the southern tip of Greenland, and then past Iceland, to encompass the entire western coast of Europe, including the complete British Isles and extending around the top of Scandinavia into the Barents Sea. Basking sharks, in addition to one species each of mackerel, lantern, and cat sharks, are all listed as being "very rare guests" in southern Greenland (Møller et al. 2010). The Egede UMO sighting, however, occurred near where the reverends were based: the community of Disko, a thousand kilometers up the west coast from the southern tip of the island. Therefore, if it was indeed a basking shark that was seen, it would have been a quite rare or anomalous sighting for the region (especially given the unfavorable climatic conditions that existed at the time—Fagan 2000). Might this novelty explain why the animal went unrecognized by the accomplished natural historians?

Greenland waters are presently thought to contain 269 species of fishes from 80 families (Møller et al. 2010). But this is an inventory that has grown steadily over the years: 1776 (ca. 28 species), 1837 (ca. 51 species), 1875 (68 species), 1926 (100 species), 1981 (116 species), and 1992 (216 species). Five of the species newly recorded between the two comprehensive surveys of 1992 and 2010 were arrivals from southern waters and presumed to be in Greenland as a result of increased water temperatures. Presence of other vagrants might simply be due to the unpredictable and idiosyncratic behavior of certain individuals. For example, a species of catshark was caught in Greenland in 1999, and again in 2000, more

Robert L. France

than seven decades after the only other confirmed report of the species; and a species of lantern shark is known from Greenland from only a single specimen caught in 1988. Although both of these elasmobranchs are small in size, that is not the case for the giant Bluefin tuna (*Thunnus thynnus*), of which there has been a single specimen caught in southern Greenland in 1900 (Møller et al. 2010). And then there is the even more intriguing case of the grey seal (*Halichoerus grypus*), a species considered to be restricted to the Labrador coast, Iceland, and northern Europe, of which an individual was photographed in a colony of harbor seals in southeast Greenland in 2009 (Rosing-Asvid et al. 2010). Local Inuit claimed to be familiar with the species but did note that it was only rarely seen. Although this is the first solid proof of the species from the island, there is anecdotal information to suggest the occasional presence of wayward individuals. Fabricius (1790, in Kapel 2005), although he did not personally observe the pinniped, describes a "long snouted seal" that was regarded as being very rare by southwestern Greenlanders. And Brown (1868) mentions both skins and a skull from the Disko Bay area which he considered to be evidence for the grey seal in the area. Although grey seals can travel considerable distances, studies reveal genetic isolation of the two populations on either side of the North Atlantic (Boskovic et al. 1996). The point here is that if seals can occasionally stray far beyond their customary geographic range, including as far north as Baffin Bay, it is not a reach to countenance the possibility that a basking shark, a much more mobile species known to regularly cross the North Atlantic, might likewise venture up into the same region. Of course, a possibility is not the same thing as a certainty. The difference is that the present supposition for the observed abnormality of the Egede UMO can be given credence through biogeographic observations as opposed to a cryptozoological belief that is contrary to the paleontological record.

Certainly the description and illustration of the observed Egede UMO match the pointed snout and mottled skin of placoid scales that are characteristic of a basking shark. And, as noted above, so does its breaching behavior. There is one more tidbit of supportive evidence for this illation. Pontoppidan (1755:181) writes that "Mr. Bing...informed his brother in law, that this creature's eyes seemed red, and like burning fire, which makes it appear it was not the common Sea-snake." Paxton et al. (2005) rightly dismiss the exaggerated hyperbole of Bing's glowing eyes description given that the sighting took place during the extended daylight of an arctic summer. But the very fact that Bing was able to discern the eyes of the UMO from a distance of no more than 15 m is perhaps itself of interest, since, when one compares online photos of the heads of breaching grey whales and basking sharks, it is much easier to discern eyes for the latter than it is for the former. Then there is one further anatomical mystery. Paxton et al. correctly comment that, if one believes Bing's illustration to be accurate, "then the strongest objection to the baleen whale interpretation of the Egede sighting is the presence of obvious teeth in the drawing" (2005:8). Other descriptions of putative sea serpents have referred to "rows of glistening teeth" and the like, traits which are similar to the papillae of leatherback turtles or the rows of gill-rakers of basking sharks (Brongersma 1968;

100

Part II: Speculations Concerning the Identity of the Egedes' Entangled UMO

France 2017; Heuvelmans 1968). The Egede UMO was obviously not a chelonian, but could the teeth shown on Bing's drawing be the oftimes easily observable gill-rakers of a basking shark? Possibly, but the sharks display these when slowly filtering plankton, and certainly not when breaching. So that train of reasoning is a dead-end. On the other hand, the account makes mention that the UMO was "3 times above the water" (P. Egede in Paxton et al. 2005:2). Unfortunately the account does not mention whether the creature was observed in-between that hat-trick of breaches. If it had been, and it was a basking shark engaged in filter-feeding, mouth prominently agape as they do, the gill-rakers would have been prominently displayed. Could this have been transferred by Bing into the single composite drawing he made of the toothed animal breaching, just as he collapsed both the temporally distinct acts of breach itself and the "tail" later being flung upward into one conjoined image? This is a complete supposition of course and strays toward to the sort of inferences that characterize, and plague, much (but not all) of cryptozoology.

It is important to reiterate that the formal scientific description and naming of the basking shark did not occur until 1765, fully three decades after the Greenland encounter, and a quarter of a century after the Egedes published their respective books. As good natural historians as they may have been, they can certainly be excused for being ignorant of basking sharks, especially if it was one, as posited here, to have been an infrequent visitor to the west coast of Greenland; notwithstanding, of course, the further anomaly of the animal breaching as well as pulling a train of entangled hunting or fishing equipment.

As a counterpoint to the preceding suggestion that the Egede UMO may have been a basking shark that was entangled, it is worth briefly considering the alternative possibility of it being a Greenland shark (*Somniosus microcephalus*) likewise so-encumbered. The biggest drawback in the basking shark hypothesis is it would have taken a considerable (but not impossible) range extension for it to be observed in central Greenland by Egede. Not so, however, with respect to the Greenland shark which occurs in abundance in the region (McNeil et al. 2012), where it is susceptible to becoming incidental bycatch (Idrobo and Berkes 2012). Also, the fact that Inuit and Norse have hunted Greenland sharks for centuries, and that a liver-oil fishery was developed during the eighteenth century to supply Europe with an illumination source, offers promise for its candidacy as the entangled UMO. In particular, from as early as the seventeenth century, the fishery in Iceland employed the use of a shark-specific sprawl-line consisting of surface floats (the Egede UMO's tail?) and a sink with baited hooks (McNeil et al. 2012). Growing to lengths of more than 6 m and rarely being seen, the Greenland shark has featured in several episodes of the hyperbolic, grandiloquent, and immensely popular television series, *River Monsters,* where the host has fished for individuals in both Greenland and Norway. Moreover, a carcass identified to be that of this type of shark that was washed up on a northern British shore led the same host to suppose that the species might be responsible for some of the sightings of the monster in Loch Ness. However, because it generally inhabits deep waters and has the slowest swimming speed and tail-beat frequency for its size recorded for all fishes (Watanable et al. 2012), it hard to reconcile this with the powerful actions required for the surface-breaching behavior of the Egede UMO. Perhaps the closest the Greenland shark comes to the Egede UMO is that due to its incredibly slow metabolism which results in it being the longest lived of all vertebrate species (exceeding 400 years; Nielson et al. 2016), there is the remarkable likelihood that there are individual Greenland sharks that are presently swimming about west coast of Greenland today which would have been alive back at the time of the 1734 Egede sighting. The oceans most certainly do harbor all manner of truly remarkable creatures without the need to invent fictitious ones.

Part iii

Natural and Cultural History

Chapter 5.

Continuing Legacy of the Egede Creature in Paleontology and Popular Culture

Interest in sea monsters peaked during the nineteenth century. Much of this was due to the contemporaneous widespread fascination held by the educated public in the spectacular discoveries then being made in the fledgling science of paleontology (France 2019a; Loxton and Prothero 2015; Paxton and Naish 2019). For many, these fossils provided the temporal glue that linked deep geologic time with the present, given that the fossils of the antediluvian creatures being uncovered could very well be the ancestors of contemporary sea monsters. In consequence, no discussion about the social or intellectual history of sea monsters can be undertaken without due consideration to nineteenth-century vertebrate paleontology (Lyons 2010; McGowan-Hartman 2013). This chapter begins by recounting how the legacy of the long tail of Egede's sea monster, reinforced by subsequent sightings of serpentine UMOs in Norway and New England (O'Neill 1999; Pontoppidan 1753, 1755), influenced the two most famous misadventures in early paleontology.

Influence of the Egede Creature's "Tail" on Nineteenth-Century Paleontology

When fossils were first found of a long, sinuous marine animal in 1832, it was misidentified as the remains of an ancient reptile and given the name *Basilosaurus* for "regal or king lizard." Later, the leading paleontologist of the day, Sir Richard Owen, coiner of the term "dinosaur" and vocal skeptic of the famous *Daedalus* Sea Serpent, correctly identified the animal as being an ancient whale, renaming it *Zeuglodon* for its double row of teeth (Figure 2.4). In 1845, in one of the most famous paleontological and cryptozoological hoaxes (or, if one is kind, blunders) of all time, the German-American fossil hunter, Albert Koch, collected bones from several such fossil whales which he then strung together and subsequently exhibited in New York as proof for the existence of a giant sea serpent. This assembly of bones was considered to resemble the famous Egede creature, with its long and slender shape corresponding to "various statements made by persons, in regard to having seen serpents in different parts of the ocean" (Koch in Rieppel 2017:139). In this regard, Koch the showman was "using the typical templates of sea serpent imagery to gain credence for his creation" (Rotschafer 2014:27).

By all accounts Koch was part competent paleontologist and part Barnumesque huckster (Rieppel 2017). He gained international renown in 1842 when he cobbled together the bones from several mastodons and elephants and exhibited the enormous skeleton in England, christening it as a new species of leviathan, *Missourium,* from its eponymous location of discovery (Jones 1989). Not being fooled, but valuing the rare bones, the British Museum bought the exhibited skeleton, which was immediately broken apart and correctly reassembled by no less personage than Dr. Owen (Rupke 1994), referred to by Prince Albert as the "sea serpent killer" due to his skepticism about such make-believe animals (Regal 2012). Emboldened by his payment of £1,300 (the equivalent of $156,000 USD today), Koch returned to America, where he soon foisted another scam upon the gullible public.

Advertised as "GIGANTIC FOSSIL REPTILE 114 feet in Length" (on an illustrated advertisement shown in Rieppel 2017), Koch charged 25 cents to see what he referred to as the fossilized remains of an extinct sea serpent, *Hydrarchos sillimanni* (the genus name meaning king of the seas, and the species name for noted Yale University professor Benjamin Silliman, a vocal believer in the existence of sea monsters). Koch artfully arranged the skeleton with its massive head elevated, and its long spinal column, to which a few ribs and parts of flippers were attached, in an undulating shape (Figure 5.1). In doing so, Koch's life-like artistic creation did much to suggest a scientific legitimacy to sightings of sea serpents (Rotschafer

Figure 5.1. Albert Koch's *Hydrarchos sillimannii* skeleton displayed in 1845 as the giant reptilian sea serpent but which, in reality, was assembled from bones cobbled together from several different individual zeuglodon fossils (in Heulevmans 1968:320).

2014). Even the positioning of the three figures standing beneath the threatening gaping jaws is no accident, for it harkens to the deep-seated fears held by many for what the seas might secretly harbor. Moreover, at a time when social standing was equated with proper behavior, the fact that it was gentlemen who were portrayed (as was also the case for the "veritable wonder" of the globster monstrosity in Figure 4.6), reinforced the implied legitimacy (Rotschafer 2014).

Never did the colossal Sea Serpent seem so real. As Loxton and Prothero (2015:228) write: "The press was as gobsmacked as the crowds. The *New York Daily Tribune* hailed it as 'indisputably the greatest wonder that ever was brought to light out of the strata which from the crust of our globe'—and a lasting monument to American greatness."

Almost immediately, however, Koch's fraud was exposed by American naturalists, many no doubt still red-faced from the debacle of three decades before in relation to the co-called baby sea serpent from Gloucester that had resulted in worldwide ridicule (Brown 1990; France 2019a). Harvard professor of anatomy, Jeffries Wyman, examined the skeleton and found its teeth to be double-rooted, indicating the bones must be those of a mammal, not a reptile (Wyman 1845). Furthermore, the vertebrae displayed different degrees of ossification, indicating that they had been sourced from multiple animals. And if that was not enough, the flipper bones were actually artificial and not natural, being created as casts made from crushed Nautilus shells. In a letter to a colleague, Wyman (1848:67) wrote "Dr. Koch, who by the way is no doctor, is a shrewd man [and] knows very well that few are sufficiently acquainted with bones to give an opinion as to the nature of the beast." Others, who were involved in the actual collection of the bones, chipped in to indicate that the fossils had in fact originated from varied locations in Alabama, and actually had belonged to several zeuglodon skeletons. Some, however, continued to believe in Koch's veracity, since the long tail of his mounted creature matched perfectly with their confirmation bias given that this was exactly how a true sea serpent *should* look, given its resemblance to the Egede, Pontoppidan, and Gloucester UMOs. The *Illustrated London News* of October 28, 1848, for example, stated that

> Professor Silliman insists the spinal column belongs to a single individual but one that is not known to science, 'although it may countenance the popular (and I believe well founded) impression of the existence in our modern seas of huge animals, to which the name of Sea-Serpent had been attached' (in Oudemans 2007[1892]:38).

When, as more evidence of Koch's deception piled up, Silliman refused to have his name associated with the suspect skeleton, Koch, nonplussed, simply substituted the name of another zoologist, Richard Harlan, whom had discovered (and incorrectly identified as a reptile) the *Basilosaurus,* and whom now, being conveniently dead, could not take umbrage. Loxton and Prothero (2015) review the evidence as to whether Koch was a deliberate charlatan or merely an incompetent bumbler, concluding it to be the former. Eventually the skeleton was bought by the King of Prussia and put on display in Berlin, allowing Koch to maintain his career of creating super-monsters (Loxton and Prothero 2015). The skeleton was destroyed by Allied bombing during the Second World War.

Sea serpents, from the original Egede UMO down to those which continue to be seen in the present-day, are of course known for their elongated tails; indeed, so much so in this regard, it could be said to be *the* defining trait. In fact, the Egede creature's long tail was so engrained in the consciousness of nineteenth-century natural science that it even influenced legitimate paleontological reconstructions. One of the most embarrassing errors in paleontology (Davidson 2002) occurred when the noted scholar Edward Cope reconstructed a plesiosaur-like reptile with the head at the wrong end. This was done for the simple reason that it seemed inconceivable that any creature could have such a long neck when everyone knew from contemporary accounts of sea monsters, such as the Egede UMO, many of which might indeed be living fossils, that such animals were characterized by the presence of long tails. It is of course my contention (see Chapter 3) that many of these long tails were in fact strings of entangled fishing, hunting, or other anthropogenic maritime debris that had become entangled around unfortunate present-day animals, none of whom were prehistoric relics.

During the nineteenth century, both natural scientists and paleontologists engaged in a frantic scramble to have their discoveries be accredited before those of their competitors. The goal, above all else, was to be able to personally ascribe a Latin binomial moniker, often with one's own surname attached, to a newly discovered species of living or extinct specimen (Ritvo 1997). The zeal with which this quest for nomenclature was pursued was the Age of Enlightenment equivalent to the medieval machinations of squabbling clerics discovering and stealing holy relics from each other's monasteries. And no story is more illustrative of this than that of the professional rivalry between Americans Edward Cope and Othniel Marsh, approaching as it frequently did over the years, to such intense animosity that it became the stuff of legend: the so-called "bone wars" (Jaffe 2000).

Considered to be a prodigy, Cope was known for both his brilliance and stubborn arrogance, as well as his sloppiness in rushing incompletely considered ideas to print too soon (Davidson 2002). In 1867–1868, he received bones sent to him by a fossil collector from Kansas, and after more than a year of work, unveiled them as the type-specimen of a new animal, *Elasmosaurus platyurus*. Unfortunately, he had attached the skull to the wrong end of the long vertebral column (Figure 5.2). This of course reversed the overall anatomy of this particular relative of plesiosaurs, which in actuality sports a long neck and short tail, to be just the opposite: a short-necked, long-tailed being. In short, a concoction that was not too dissimilar to the Egedes' prototypical sea monster, as displayed in the well-known illustration by Reverend Bing.

Cope gave presentations to the American Philosophical Society in the fall and spring of 1868–1869, and arranged to have a preprint of his article circulated during summer 1869. The following spring, Cope's mentor, Joseph Leidy, published an article making note of Cope's er-

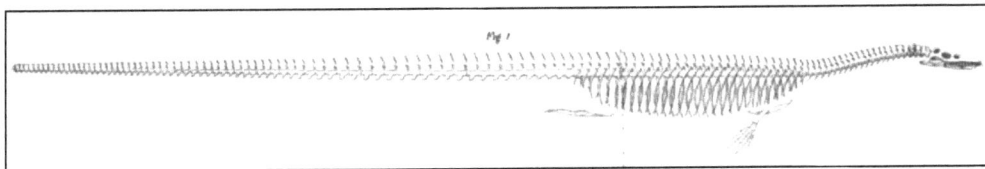

Figure 5.2. Edward Cope's "bone-headed mistake" (*sensu* Davidson 2002), the biggest blunder in nineteenth-century paleontology, which had him, in 1868, attach the head of his plesiosaur to the wrong end of the skeleton; i.e., at the terminus of its tail rather than its neck (Davidson 2002:16).

ror. This sent Cope into a flurry in attempt to recall and gather up all copies of the original preprint and its embarrassing illustrations, while at the same time, correcting the mistakes by switching the words "neck" and "tail," and redrawing the skeleton (Davidson 2002). Unfortunately, he was not able to completely suppress his blunder, as a copy of the original preprint had found its way into the hands of his arch-nemesis, Marsh, whom repeatedly over the years ahead would never let an opportunity go by without bringing it up to humiliate his foe. The one exception to this display of antagonism was when Marsh named, in a seemingly uncharacteristic burst of generosity, a new species *Mosasaurus copeanus*, after his enemy. Mosasaurs were indeed impressive creatures (see Appendix 1.3). However, as Ellis (2003) notes, this honor is much less magnanimous than it at first seems, given the conjoining together of the words "Cope" with "anus."

Davidson's 2002 paper is based on investigating the puzzlement as to why such a competent paleontologist as Cope could have made such a "boneheaded blunder." She dismisses the idea that he would not have known better due to his immaturity and lack of formal education, given the world-wide knowledge about the correct anatomy of plesiosaurs at that time; i.e., that their most distinguishing characteristic was their improbable but real elongated necks (Figure 5.3; see Appendix 1.5 and 1.6). This makes "his mistake with *E. platyurus* all the more surprising" (Davidson 2002:219). After going through all the evidence, Davidson concludes that

> put simply, Edward Cope should have known to place the skull of *E. platyurus* correctly because plesiosaur vertebral columns and the backbone of the elasmosaur have vertebrae which point in the direction of the tail, not of the skull. That is what Leidy told him (2002:218).

But still Cope persisted in getting things back-to-front. Davidson remains puzzled: "So the question again must be asked how could Edward Cope have incorrectly reconstructed *E. platyurus*" (2002:226)? It is even more baffling to her that he actually made the same error more than once.

As communicated in a letter to his father concerning the arrival of the bones, Cope was wrong from the very get-go: "It possess a tail of great length which was elevated compressed and adapted for sculling the ponderous body through the water. The limbs appear to have

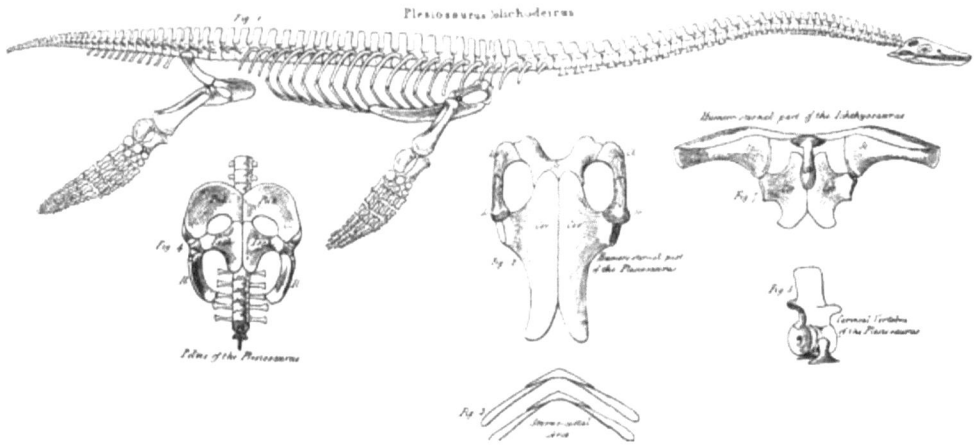

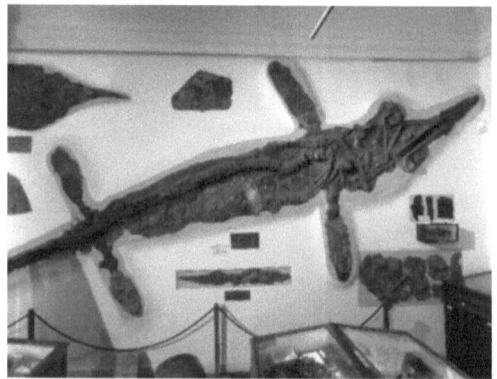

Figure 5.3. Distinctions between prehistoric "sea dragons." The upper illustration is of one of the earliest assembled plesiosaur skeletons, as described by Connybeare (1824:385). Photographs are from the Whitby Museum in Yorkshire, England. The middle image shows a long-necked plesiosaur (lower skeleton) and a short-necked, long-tailed ichthyosaur (upper skeleton), and the bottom two images the same for another ichthyosaur. See also photographs in Appendix 1.2 showing a short-necked pliosaur, a relative of the plesiosaur, which demonstrate how confusion might occur.

been disproportionally small" (in Davidson 2002:222). In short, as she writes, Cope had "in effect disassembled and reassembled the fossils into how he thought the animals should look. And that reconstruction included an *Elasmosaurus* with a short neck" (Davidson 2002:226). My contention is that the resulting short neck was merely a secondary byproduct that had to result, *ipso facto*, from where Cope's true focus really lay: that at the other end of the skeleton, and his desire to have it conform to the established tradition of long-tailed sea serpents, dating back to Egede.

The answer to Davidson's question, therefore, is that Cope was so preoccupied with sea serpents that his confirmation bias simply would not allow him to countenance ancient reptiles that looked different from their descendants that were probably still swimming about, occultly, in present-day oceans. Indeed, Cope developed ideas about ancient marine reptiles that were idiosyncratic, classifying them into a new order due to their imagined snake-like bodies (Ellis 2003). With respect to mosasaurs, Cope (1869a:184) stated that

> We may look upon the mosasaurs and their allies as a race of gigantic, marine, serpent-like reptiles...Adding a pair of short anterior paddles, they are not badly represented by old Pontoppidan's figure of a sea serpent. [by which he actually means Bing's original drawing of Egede's monster as reprinted in Pontoppidan]...Thus in the mosasaurids, we almost realize the fictions of snake-like dragons and seas serpents, which men have been ever prone to indulge.

It is my contention that Cope's use of the word "fictions" is somewhat disingenuous as I believe him to have personally harbored belief in the existence of sea dragons sporting long tails, *à la* the Egede UMO. For in the same year he published an article intended for a more general readership, in which he writes:

> While grim and monstrous Dinosaurs ranged the forests and flats of the coast of the cretaceous sea, and myriad's of Gavials [i.e., large crocodiles] basked on the bars and hugs the shores, other races people the waters. The gigantic Mosasurus, the longest of known reptiles, had few rivals in the ocean. These Pythonmorphs [a term of his own coining to indicate snake-like qualities] were the sea-serpents of that age, and their snaky forms and gaping jaws rest on better evidence than he of Nahant can yet produce (Cope 1869b:84).

The reference to "Nahant" here is significant, as it refers to the peninsula of that name in Boston Harbor, the second home of the Gloucester Sea Serpent of 1817–1819, a location where cruises were run and a hotel built to cater to the tourism industry that developed focused on providing opportunities to catch a glimpse of the famous UMO (Burns 2014; France 2019a). So once again, Cope seems, on one hand, to scoff at recent sightings of sea serpents while, on the other, to forcefully construe paleontological evidence to fit his belief in the serpentine anatomy of extinct animals. This confusion is evident in the plate that accompanied the article (Figure 5.4), wherein, the artist, working from Cope's sketches, depicts a short-necked *Elasmosaurus* in addition to a *Mosasaurus* that resembles a plesiosaur, both shown, in homage to Egede and Pontoppidan, with elongated tails of the classic Nordic sea serpent. It is possible that Cope may have also been thinking of crocodile-like, short-necked pliosaurs (see Appendix 1.2 and 1.3) when he drew the original mashup for the artist to work from.

Davidson (2002:225) certainly hits the mark when, in exasperation, she writes that "Something is frightfully confused in this article," noting that by this time, 1869, Cope had almost certainly been informed of his error by his mentor and friend Leidy, but then goes on to make the same mistake again. She believes Cope based his mosasaur illustration upon two plesiosaurs shown in the frontispiece engraving in Thomas Hawkins' 1840 *The Book of the Great Sea Dragons*. If that is so, it is possible that the inspiration for the enormously long-tailed plesiosaurs in Hawkins' book extends further back to Bing's rendering of the Egede sea

COPE ON FOSSIL REPTILES OF NEW JERSEY.

Provided by: http://www.strangescience.net
Originally appeared in: "Fossil Reptiles of New Jersey" in *American Naturalist*
Now appears in: *The Dinosaur Papers* by Weishampel and White

Figure 5.4. Cope's continuing confusion, in which his depiction of the prehistoric fauna of New Jersey continues the mistake of showing a short-necked, long-tailed plesiosaur as well as a mosasaur that looks like a plesiosaur (Cope 1869b:87).

monster. This then might be the true terminus to what Davidson refers to as "the tale of the tail com[ing] to its end" (2002:236).

Legacy of the Egede Creature in Sea Monster Lore and Cryptozoology

Both preceding examples from the early days of paleontology demonstrate the enduring legacy of the famous Egede sighting in terms of how we envision the largest, either real or imaginary, denizens of the deep. Children are fascinated with prehistoric animals like dinosaurs because they are big and scary-looking without being truly threatening since they are known to no longer exist (Schowalter 1979). Cryptozoologists are fascinated with prehistoric animals like predaceous whales and aquatic reptiles because they are big and scary-looking and are mysterious due to the imagined possibly that they might still exist (see Coleman and Huyghe 2003).

During the great heyday of natural history in the nineteenth century, when sea monsters were a legitimate field of discourse in the scientific literature (Lyons 2010; Westrum 1979), speculations about new sightings of UMOs would often make reference to the Egedes' mysterious creature. In this respect, Heuvelmans (1968) was correct in his belief that the Egede encounter cast a large shadow of influence over subsequent sightings of sea monsters. Four examples will suffice. Samuel Taylor Coleridge, certainly no stranger himself to mysterious maritime lore (Ower 2001), served as secretary for the noted sailor and governor of Malta, Sir Alexander Ball, and communicated the latter's reflections on sea serpents. According to Coleridge (in Heuvelmans 1968:116), Rear-Admiral Ball, reticent of the "general opinion [of those doubting] the relations of Egede," kept mum for years about his own 1781 encounter in the Baltic Sea with a "creature resembling in all respects the Monster described by Egede under the name Sea Serpent." Next, Heuvelmans (1968:246) summed up a handful of nineteenth-century sightings from around the world with the statement: "In short, in these five cases the creature is either a maned sea-serpent, or a 'super otter' with a short neck and long tail, like that so well described and pictured by Hans Egede." Then, in another incident, a 30 m long "immense sea monster" seen in 1849 off the coast of Florida, raised its head high above the water, displaying a tapered neck and its body in the process, the latter revealing a set of meter-long front appendages. Oudemans (2007[1892]:228) has no doubt what was seen: "At a glance we recognize the sea-serpent, as it appeared to Hans Egede. 'The largest portion of its body' was seen, 'and a pair of frightful fins or claws, several feet in length.'" Oudemans follows by directing readers to go back and compare this description to the Egede creature shown in Bing's illustration presented near the beginning of his book. Heuvelmans (1968:231), however, is less certain that the two UMOs were of the same typology:

> The depiction of the creature is plausible enough—it reminds one of Hans
> Egede's 'most dreadful Monster'—but its movements are less so. It is hard

to imagine how it could lift 'the largest portion of its body' out of the water except in a sudden leap or somersault.

He then cautiously concludes that "one cannot help wondering whether it may not derive from an old picture of Egede's monster rather than something actually seen." And finally, there is the 1891 sighting of an UMO in New Zealand which Gould (1930:17) presents due to the similarity of the creature to that of Egede:

> It would from time to time lift its head and part of its body to a great height perpendicularly, and when in that positon would turn its body round in a most peculiar manner, displaying a black, white belly, and two armlet appendages of great length, which appeared to dangle about like a broken limb on a human being. It would suddenly drop back into the water ...

The original anecdote actually continues with the "appendages" described as being sent "scattering in all directions" during the animal's plunge, a fact which led me (France 2020a) to conclude that the appendages were really trains of entangled debris, and not, as Gould would have, fore-flippers. So, this UMO is indeed similar to that of Egede's, but perhaps for a different reason than originally proposed.

Early natural history was infused with a strong theological underpinning (Barber 1980; Berger 1983). As previously mentioned, one of the big conundrums that the devout had concerning paleontology and evolutionary biology in the nineteenth century was their difficulty in accepting that the Creator would populate a world with creatures which would become extinct before humans ever arrived on the scene. This led to a widespread belief that the giant creatures represented in fossils must still be in existence somewhere else in the remote, and as then still incompletely-explored, world. As early as 1846, naturalists were opining as to whether representatives of Koch's *Hydrachus* still swam in the seas. Later, Oudemans (2007[1892]) posited on whether zeuglodons and sea serpents were one and the same. And more recently, Heuvelmans (1968) championed this particular argument, linking it all back to the Egede UMO which, as mentioned, can be said to be the foundation for the modern sea monster tradition:

> Might not the zeuglodon ... make a very satisfactory sea-serpent? Some species of zeuglodon were more than 60 feet long. They were exceptionally slender animals and much more serpentine than the cetaceans of today. Instead of a tail spreading sideways in flukes, they had a long tapering one ending in a point ... Like other cetaceans they have only the front limbs, in the form of flippers, but they were more flexible, for the upper part of the arm was not so atrophied, and the fingers were still separate and probably

had claws. Their skeletons also reveal that, unlike our whales and dolphins, they did not have vents on the top of the head, but nostrils in the usual place at the end of the snout. They had even longer teeth than dolphins…The long cervical vertebrae, articulated like a seal's, enabled them to turn their heads freely on a neck, which, though relatively short, was loose and flexible.

There is one big snag to the zeuglodon theory. It cannot explain those sea-serpents which have heads and necks [shaped] like an umbrella-handle or a periscope, for the zeuglodon's neck is too short. But 'that most dreadful Monster' seen by Egede is a very different matter. *One could hardly wish for a better reconstruction of a zeuglodon than Bing's drawing*. The general shape, the long head with nostrils on the nose, the whale's breath, the flexible neck, the long tapering tail, the single pair of flippers—it all agrees. Yet Bing drew his picture a hundred years before the first fossil zeuglodon bones had been discovered or anyone could possibly guess that such an animal might exist (Heuvelmans 1968:185–186, emphasis added).

Heuvelmans goes on to state that it would not be surprising if zeuglodons were to survive given that they existed 30 million years ago, which, he informs, trotting out the same cliché and flawed non sequitur logic of many a cryptozoologist, is half the length of time that coelacanths (*Latimeria* spp.) survived unnoticed. He concludes by expressing surprise that nobody took any immediate notice of the 1846 zeuglodon theory. But cryptozoologists have certainly made up for that initial omission in a big way ever since. As many skeptics have noted, there is an obvious faddism in the cryptozoological ascription of candidate animals to explain sea monster/serpent sightings. Today, the most popular cryptid is the zeuglodon, as witness to the fact that it is now the dominant theory put forward for the Egede UMO (Table 3.1). And much of this is based on Heuvelman's reasoning that "one could hardly wish for a better reconstruction of a zeuglodon than Bing's drawing" (1968:186).

The *Illustrated London News* of November 1848 shows how Bing's illustration, conjoined with drawings of the *Daedalus* sea serpent (Galbreath 2015), and the deformed snake purported to have been a baby of the Gloucester UMO (France 2019a), to form a tryptic of the most famous nineteenth-century marine mystery animals; a sort of homage to the holy trinity of cryptozoological belief. No surprise then that Egede's "most dreadful monster" was repeatedly dusted off the shelf to enjoy an active afterlife through being bandied about in discussions in the popular press following the *Daedalus* affair of 1848 (which became the most debated sea monster sighting of all time). Similarly, following the increasing popularity of Lee's (1883) giant squid theory, once again the Egede UMO was shown, along with some of the usual suspects (Figure 5.5), in *Sea Monsters*, a book for young adults (Beard 1887). Fiction also made use of the Egede creature in a like fashion. For example, Jules Verne's (1870)

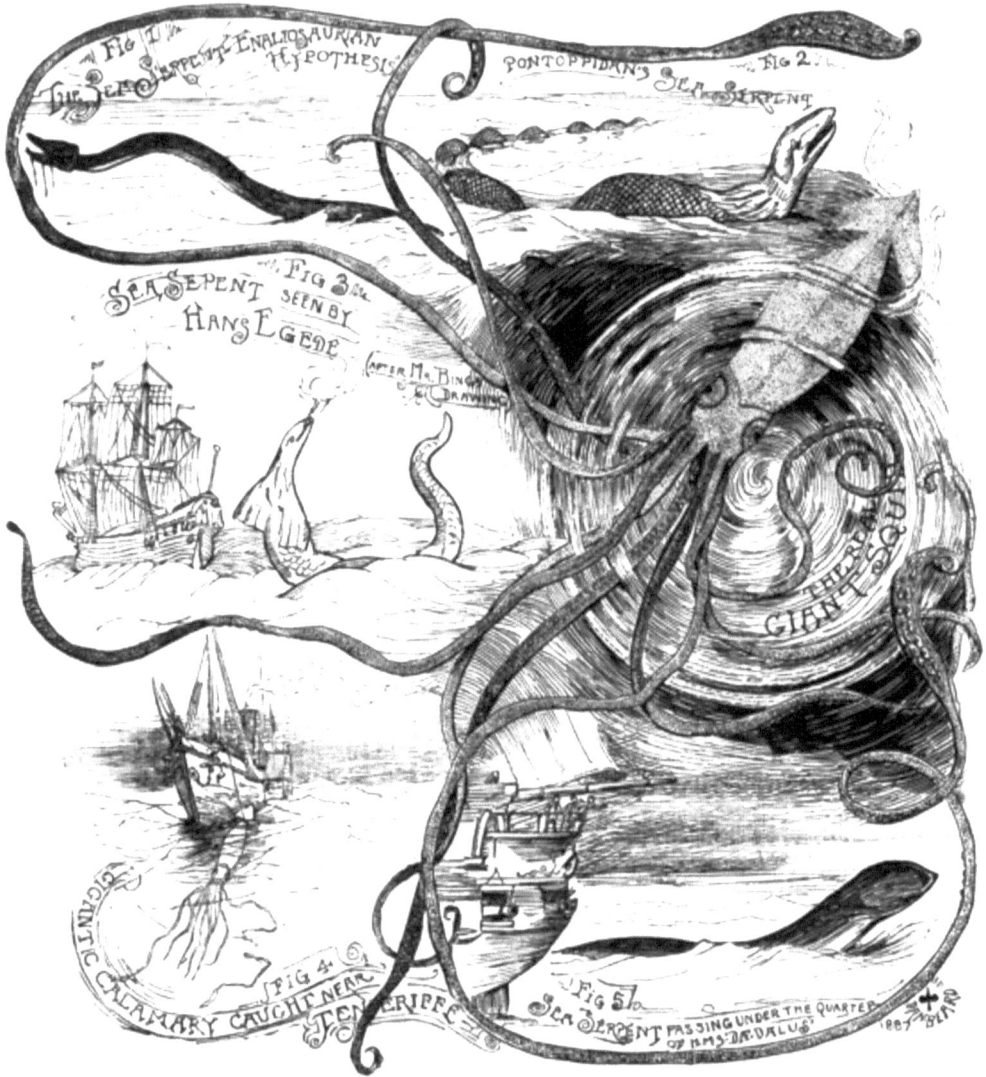

Figure 5.5. The Egede UMO joins other famous "sea monsters" in an illustration from Beard (1887:192) that implies that all such can be explained by Lee's (1883) giant squid theory.

classic novel, *Twenty Thousand Leagues Under the Sea,* begins with a review of the worldwide attention brought about by repeated sightings of a mysterious sea monster (which of course would turn out to be the submarine *Nautilus*):

> In every large centre of population, the monster was all the rage. Songs were swung about it in cafes, journalists mocked the very ideas that such a thing existed, and plays featuring it were staged in theatres…When newspapers were short of copy, they resurrected all the old tales of legendary, gigantic creatures … Reports dating from ancient times were reprinted, the

> pronouncements of Aristotle and Pliny, who believed in the existence of monsters, the Norwegian accounts of the Bishop Pontoppidan, the travels of Paul Egede and, not least, the reports of Mr. Harrington [who had observed another well publicized UMO in 1857] … (Verne 1870:7).

Similar referrals to the Egede encounter frequently took place within the scientific literature as well, as for example, in an 1880 *Nature* article, wherein Searles Wood suggests that most sea monster/serpent sightings are of zeuglodons, although unlike Heuvelmans he seems not so sure when it comes to Egede's UMO:

> It is most likely that Bishop Pontoppidan…concocted his two figures (one of which is that of a huge snake undulating on the waves, and the other [i.e., his copy of Bing's illustration] that of a serpent-like animal with pectoral flappers or fins, resting almost on the surface of the sea, with head and tail erect out of the water like the letter U, and spouting water or steam from its mouth <u>in a single column</u>), from accounts given him by Norwegian seamen, some of whom had seen the animal in the position in which it was observed from the *Daedalus* (Wood 1880:812).

Oudemans takes issue with some of Wood's statements, such as that the Egede animal was "resting" on the surface in that giant U-shape, when it was clearly "seen in this position for only the fraction of a second!" (2007[1892]:332). Particularly, he goes on to comment upon the exhalation shown in Bing's illustration:

> Mr. Wood, describing the drawing of Mr. Bing underlines the words: *in a single column*, speaking of the animal's 'spouting water or steam from its mouth.' Now I ask my readers (drawing their attention to the fact that the figure represents the animal's head seen from aside), whether a column, spouted from the animal's nose or mouth, when seen from aside could ever have been decided to be single or double! If we look at the breath of a horse, standing just on one side of him, it will be observed to be single. This optical illusion will be dispelled as soon as we stand in front of the horse. Bing's figure would have been incorrect, if he had drawn two columns, though in reality—if the animal exhaled through its nostrils,—the column must have been double (Oudemans 2007[1892]:332, emphasis in original).

Oudemans makes much ado about all this as it is necessary for advancing his own pet theory that the Egede animal was a mammal breathing through a *pair* of nostrils and not some type of prehistoric reptile spouting a single stream of water or steam from out of its mouth.

For today's cryptozoologists, there are few historical sightings of sea monsters that are of such talismanic importance as the one described by the Egedes. One example will suffice. In a 1991 paper published in a fringe-science journal, Michael Swords, a self-professed believer in UFOs and paranormal phenomena, undertook an anthropological review of the folkloric beliefs in strange sea animals held by Indigenous peoples of coastal British Columbia. So far, so good. However, Swords deliberately disregards cautionary statements made by folklorists Meurger and Gagnon (1988) not to confuse shamanic metaphors with actual historical precedents, and goes on to associate the former with a modern sea monster thought to reside in the region (not surprisingly, non-extinct zeuglodons factor into the story). Along the way, he mentions a small Aboriginal ivory carving in the Chicago Field Museum's collections that is labelled "sea monster," writing that:

> Indeed, it seems a good candidate for our quarry, and it lies alongside several other artifacts, easily identified [assumedly he means of other known marine life], attesting to the realistic craft skills of the people. It is pictured as aquatic, elongated, toothed, snouted, short-horned, big-eyed, fore-flippered—in fact, a perfectly good Sisuti-Wasgo [i.e., the "Sea-Wolf" in the regional Native mythology] in a more zoologically credible artistic rendering than most of the wonderfully exotic artifacts of the area (Swords 1991:91).

At which point, Swords continues: "As an amusing, perhaps meaningless, aside, the creature is arched, head and tail rearing, in a similar manner to Hans Egede's sea serpent of 1734 in the Atlantic arctic" (1991:92). He then concludes by showing his own renderings of both the Indigenous and Egede monsters (Figure 5.6). Despite his mollifying statement, such evidence is in fact used later in the article to give credence to the possible existence of the region's present-day cryptid (LeBlond and Bousfield 1995). Similarities drawn between contemporary

Fig. 6. Comparison of the Chicago Field Museum of Natural History's Tlingit "Sea-Monster' with Hans Egede's rearing sea-serpent of 1734. (redrawn by author).

Figure 5.6. Cryptozoologist Michael Sword's (1991:92) likening of a First Nation's ivory carving of a mythical creature from British Columbia, Canada with the Egede UMO from Greenland.

bones or sightings and antecedent descriptions of mythological animals is a feature of early paleontology (Mavor 2000; McGowan-Hartman 2013), just as it is of modern cryptozoology (Meurger and Gagnon 1988; Loxton and Prothero 2015).

As well as being a favorite poster-child in hundreds of cryptozoology websites, Bing's illustration continues to be trotted out in the popular media to provide a sense of historical verisimilitude. Two examples suffice. Firstly, in the glossy book accompanying a 2010 museum exhibition in Saguenay, Quebec, called *Fantastic Sea Monsters* (Cazeils 2011), a full page is devoted to Pontoppidan's fusion illustration showing the incorrectly portrayed Greenland encounter. In this case, an error is perpetuated by ascribing the drawing to Egede's 1741 book. Secondly, the cover masthead of a magazine entitled *Mysteries of the Loch Ness Monster* unabashedly announces that it is produced "From the Secret Files of the National Enquirer" (Americana Media Specials 2019). The richly illustrated publication begins with a full-page (this time non-ascribed) reprinting of Pontoppidan's version (Figure 5.7). This would neither be the first nor last time that the Egede creature would be associatively transposed from its Greenland marine home to a distant lacustrine environment. Most recently, during the flurry of international reportage in late 2019 concerning a biodiversity survey of DNA in Loch Ness—which finally and conclusively laid to rest the notion that there had ever been a population therein of resident plesiosaurs—many television news stations ran a montage of putative illustrations or photographs of famous aquatic cryptids to accompany the story (i.e., this was presumably due to DNA strands not being dramatically photogenic enough). And there, in sequence juxtaposed to the famous so-called "surgeon's photo" of Nessie-as-a-plesiosaur, was James Stewart's fanciful version (Figure 1.12, cover) of Bing's original illustration. That an image from Hamilton's book on aquatic natural history published in Victorian Britain can find resonance 180 years later on North American cable news stations, shows the lasting importance of the Egede UMO on modern consciousness. The next section goes on to explore one particular aspect of this enduring legacy.

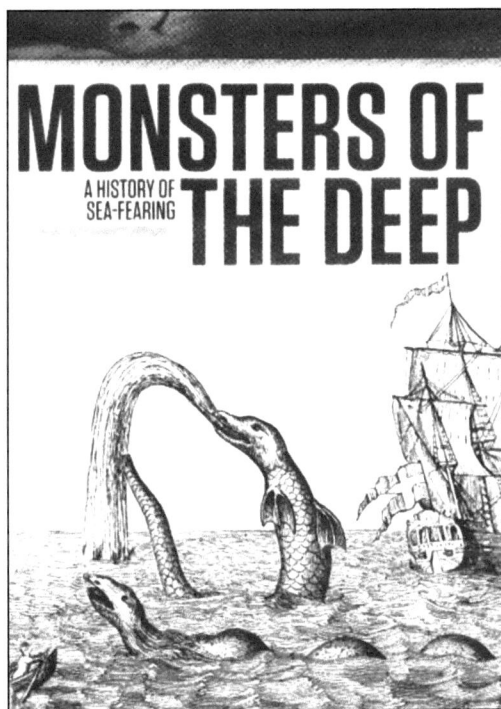

Figure 5.7. Interlinking of the Egede and Pontoppidan UMOs with "Nessie," the fictional Loch Ness Monster, in a popular magazine about the latter (Americana Media Specials 2019).

Artistic Imagination Inspired by the Egede Creature's Breaching Leap

Because cryptozoology inhabits a make-believe world, it roots lie in fiction rather than in fact. In consequence, monsters "originate not in the backwoods of the world but in the minds of imaginative individuals" (Mullis 2019), often expressed as folkloric belief (Meurger and Gagnon 1988). It is for this reason, therefore, that the over-earnest endeavors of cryptozoology's adherents can be so easily parodied, not only by scientists (Sheldon and Kerr 1972), but also by artists, as for example, in theater (Crafts 1819), film (Penn and Herzog 2004), and fictive art (Robertson 2020). Paleontology's role in fueling cryptozoological fancy has been noted (Paxton and Naish 2019). But this is often mediated by the important intermediary stage in terms of the influence of science on the perception of a prehistoric world created through art, including sculpture (Rupke 1994; see Appendix 1.5), illustration (Rotschafer 2014; Rudwick 1992), fiction (Angeno and Khouri 1981; Mullis 2019), and film (Jylkka 2018; Loxton and Prothero 2015). In consequence, cryptozoology can, and perhaps primarily should be, interpreted through an artistic lens. As Rotschafer (2014:Abstract) states:

> Art infused historically specific meaning into the schematized and persistent form of the sea serpent. It is not only that people saw serpents that looked like the ones they had seen in pictures, but that the pictures offered a kind of template on which viewers could inscribe particular historical fears.

As a dramatic vehicle in which to startle readers and viewers, writers and artists have been inspired by the breaching whale/basking shark aspect of the Egede encounter (which, as previously mentioned, may have had its origin in Ortelius' Icelandic *Staukul* monster). One notable example is the computer-generated image of a leaping *Mosasaurus* (Figure 5.8) created for the 2004 BBC show *Chased by Sea Monsters* (see Appendix 1.8) and used as the opening image in the accompanying book (Marven and James 2004). In this case there is no denying, embattled pteranodon notwithstanding, the striking resemblance of the prehistoric predator to the UMO shown in Bing's 1734 illustration. This motif was revisited a decade later in three dramatic scenes in the film *Jurassic World* (O'Byrne 2018) wherein a similar *Mosasaurus* erupts from out of the water to grab a great white shark dangled over the Sea World-like enclosure as bait (Figure 5.9), a scene replicated through use of a giant model constructed in London to advertise the United Kingdom release of the film (Figure 5.9).

One particular aspect of the Egede encounter has inspired writers of terror and fantasy. It concerns the question that, what if following its breaching, the "very horrible sea-creature" did *not* narrowly miss the ship and plunge back down into the depths, waving its tail about in the air as a too-da-loo gesture, but instead careened down right onto the deck? Certainly the possibility seems plausible if one is to believe all those illustrations from old maps and books that depict giant sea serpents draped over ships while snacking upon hapless sailors (Fig-

Figure 5.8. Striking a very Ortelius *Staukul* monster or Egede UMO-like pose, a leaping mosasaur grabs a quick snack on the fly in Marven and James (2004:125) book *Chased by Sea Monsters,* which derives from the meritorious BBC series of the same name (see Appendix 1.8).

Figure 5.9. The first Egede UMO-like leap of the *Mososaurus* in the film *Jurassic World* (Public Domain: see Photo Credits page), and an advertising model of the scene which was temporarily erected Thameside in London (WEHN Rights Ltd./Alamay Stock Photo).

ure 5.10). For did not Magnus himself state: "A single one of these monsters can quite easily capsize or sink several large ships crammed with the strongest sailors," and that those beasts which were as "huge as mountains" would accomplish this "if they are not frightened away by the sounds of trumpets or by throwing empty barrels into the sea" (Nigg 2013:150; Szabo 2008:200; Figure 5.11). In this mindset, it is terrifying to contemplate just how lucky Egede and company were to escape a similar fate. As a result, the corpus of sea monster sightings contains examples of purple prose wherein vessels and mariners were similarly fortunate. Three such from Paul Harrison's *Sea Serpents and Lake Monsters of the British Isles* (2001) will suffice. "The huge body of the monster was towering as high as our masthead, and the hissing sound it made struck terror to our hearts, as we feared that every moment it might make a dash forward carrying the sloop over" (Harrison 2001:191; from 1898); and

> I, who had been four years to the Arctic regions and has sailed on nearly every coast in the known world, had seen many strange phenomena at sea, but we were about to witness a sight the like of which I, or any of my mates, had never dreamt of…Towering in the air a dark looking object, possibly 300 feet in height, and with a long projecting head and two fore fins or legs making a steady swimming motion and emitting a hissing sort of sound, was now quite clearly seen, the eyes, the mouth, and even the nostrils were plainly in view: and all the prehistoric monsters I ever read of were mere mites compared with this mammoth. What eyes! The sight of which seemed to make everyone spellbound and powerless, a row of teeth, each one four times the length of a man in appearance, but of so terrifying an aspect that it was a few minutes before we realized our proximity to the monster. The

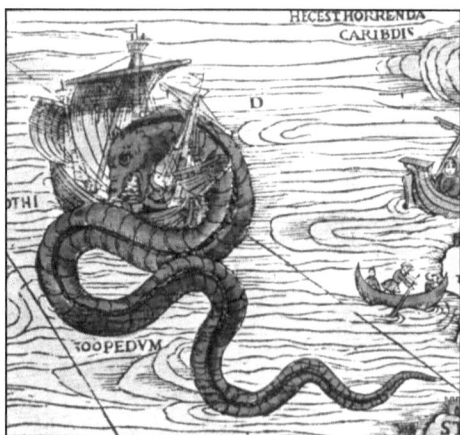

Figure 5.10. A close-up of the sea serpent illustrated on Magnus' *Carta marina* (in Nigg 2013:13) and portrayed in the act of wreaking havoc by draping itself over a ship while sampling sailor *hors d'oeuvres*.

Figure 5.11. Another close-up from the *Carta marina*, (in Nigg 2013:13) this one depicting the protection strategies of hurling barrels into the water and tooting on trumpets to distract malevolent monsters.

putting the boat up in the wind brought us to our senses, and all hands sprang to put the boat about – the audible prayers of the hired hand being the only words spoken, we all felt so impressed. The face of the old man looked like a sheet of white canvas, and terror seemed depicted on every face. We all huddled together after putting about, and no one said a word, nor dare we look behind; but, still hearing the lashing of the water and the hissing sound, we knew we were yet within a close distance of the monster, if it were not really following us. The mantle of darkness, however, was in our favor, assuring us that we were now out of sight and reach of it. Never did any crew reach a port more thankful than we did, and as each winded their way home their thoughts were, 'What could it be? Will anyone believe us?' Yes it is a verity, and the whole truth of our crew can prove it, and that this is the exact truth. (Harrison 2001:191–192; from 1891).

And in another example of hyperbolic prose that reads as if the writer were emulating a Victorian gothic thriller (Harrison 2001:160; from 1871), wherein the UMO was probably, as also the case for the previous sightings, either a breaching whale or basking shark, or a whale simply elevating its head for spy-hopping, resembling, therefore, the Icelandic *Staukul* monster:

It was dusk, the sky was still light, but the land was dark – a fine night with a light northerly breeze and a ripple on the water, Sandy and the two old men began to haul their net. He was only a young boy and his arms tired easily. He rested for a moment and as he did so, he noticed an object rising out of the water 50 yd to seaward of them. It was about yard high when he first saw it, but as he watched, it rose slowly from the surface to a height of 20 or more ft – a tapering column that moved to and fro in the air. Sandy called excitedly to the old men, but at first got only an angry retort to keep hauling the net and not be wasting time. At last Stewart looked up in exasperation, and then sprang to his feet in bewildered astonishment, as he too saw what Sandy was looking at. While this 'tail' was still waving in the air, they could see the water rippling against a dark mass below it which was just breaking the surface, and which they presumed to be the animal's body. The high column descended slowly into the sea as it had risen and as the last of it submerged the boat began to rock on a commotion of water like the wake of a passing steamer. The three occupants of the boat elected to leave the area as quickly as possible, dropped their nets and rowed for the safety of the shore.

From *shokushu goukan* (tentacle erotica) to beachside thrillers, there has been a recent burgeoning of popular fiction involving sea monsters, to such a degree as to merit academic study (Hackett and Harrington 2018). A good number of these novels use the tweaking of the famous Egede encounter by having the sea monster rise up out the waters followed by crashing upon, and sometimes sinking, the ship. These are too numerous to consider here in the present context. Instead, as an example of this aspect of the Egede-UMO legacy, I will go back to consider one of the very first instances of its inspiration on creative writing.

Eugene Batchelder's 1849 *Romance of the Sea Serpent, or, the Icthyosaurus,* may be the earliest novel to be focused on a single mysterious denizen of the deep. As the case for the contemporaneously published *Moby-Dick* (Post-Lauria 1990), it too is regarded as a "mixed-form novel" wherein factual events are used as the storyline source. In this respect, the novel is really a work of grotesque fantasy, not fiction per se (Burns 2014). Interestingly, in his compendium of sea serpent sightings, Batchelder mentions the 1825 Halifax, Nova Scotia encounter, which was suggested, even at the time, to have been an entangled basking shark (Hebda 2015). Batchelder's fictitious sea serpent is inspired by the one that visited Glouces-ter, Massachusetts during 1817–1825 (France 2019a, 2019b). The protagonists encounter a Nordic mariner who tells of an encounter his ship had with "a savage beast, [that] leaped o'er between her masts, and sunk her … The ship was lost, and only Joe escaped these Ormens[8] of the Soe" (Batchelder 1849:49). Later, off the Nahant Peninsula in Boston Harbor, the loca-tion of many sea serpent sightings (France 2019a), the mariner describes a similar "horrible monster," that leaps, Egede UMO-like, out of the water, this time to wreck the small boat and then consume its crew (Figure 5.12). But it is that footnote "8" that is of most interest, for it leads one to the expansive series of notes (which actually comprise 73 of the total number of 191 pages of the book) which Batchelder uses to support his grotesque fantasy. For here, he specifically cites Egede and makes reference to Miguel de Cervantes' Byzantine romance novel *The Travails of Persiles and Sigismunda*, completed just before the great novelist's death, which was posthumously published in 1617, and which, though little read today, the author considered to be his crowning achievement (even over *Don Quixote*).

Batchelder begins by explaining that "Ormens of the Soe" is a Nordic expression mean-ing snakes of the sea. He then mentions that in response to the flurry of sightings of the Gloucester Sea Serpent in 1817, the Linnaean Society of New England solicited corroborat-ing evidence, one such testimonial relating to events that had transpired earlier in the state of Maine. He then quotes this particular account, which reads like so many others as to strain credulity:

> [He] declared that he had often seen a marine monster of this description, which was as large as a sloop's boom, about sixty or seventy feet long. He as-serted, that, about the year 1780, as a schooner was lying at the mouth of the river, or in the bay, one of these enormous creatures leaped over it between the masts; the men ran into the hold for fright, and the weight of the serpent sunk the vessel, which was of eighteen tons burthen (Batchelder 1849:132).

Figure 5.12. The fate of unfortunate individuals in Eugene Batchelder's (1849:142) *Romance of the Sea Serpent, or, the Icthyosaurus,* whom obviously did not heed Magnus's advice to carry decoy barrels and a trumpet when plying waters infested with sea monsters.

Next, Batchelder moves back in time by referring to Cervantes:

> Now it is a singular fact, that the Spanish sailors, and perhaps the Spanish naturalists, as early as the year 1617, just two hundred years before the Linnaean Society received this statement... were aware of the fact that there were sea-serpents on the coast of Norway; and it would seem that they were aware that they [the sea serpents] sometimes came on board in a rather unceremonious manner. At all events, Cervantes... gives this terrible account of the Sea-Serpent in the last romance he ever wrote... The scene in which the Sea-Serpent is introduced is laid in the North Sea, off the coast of Norway; and it is not to be presumed that a great writer like Cervantes, with his usual fidelity to Nature, would have introduced into what he intended should be his great work this story; unless he had for it some foundation; still if this is purely his own invention, it is certainly a wonderful coincidence that the scene is laid on the coast of Norway, where Pontoppidan asserts, just such attacks have happened, and where the Rev. Mr. Egede saw, as he declares, 'on the sixth of July, 1734, a very large and frightful sea monster, which raised itself so high out of the water...' (1849:133).

127

Robert L. France

So here we have the first linking together of the words "foundation" and "Egede" as used in the title of the present monograph. Batchelder continues by providing the entire Egede account, citing the latter's *Journal of the Greenland Mission*, from which he then segues to quoting from Cervantes' novel:

> I was sitting on the deck, when suddenly it began to rain, not drops, *but whole sheets of water upon the ship, in a manner* that appeared *as if the sea had been taken into the air and fallen upon the ship*. All suddenly arose, and looked on every side, but we saw *the heavens clear*, and no signs of a hurricane; and those who were with me said 'Without doubt this rain does not come from the heavens, but from the heads of *those monstrous fishes they call shipwreckers*; and, if so, we are in *great danger of being lost...At this, we saw raised and put into the ship* a NECK like that of a TERRIBLE SERPENT, *who* TOOK OFF A MARINER, and *swallowed him* quickly, without so much as even chewing him! This was done amid the confused noise of the mariners, who did not dare to rise on their feet, for fear of being carried off by this horrid monster (Batchelder 1849:134, emphasis in original).

The crew fires off a cannon and put all sails to the wind to effect their escape. Batchelder finishes his footnote "8" with long passages from Pontoppidan based on eyewitness accounts of UMOs that similarly sprang from out of the water and threw themselves over boats.

Despite Batchelder's geographic confusion that has him placing Greenland beside Norway, it is interesting to consider that Cervantes' "horrid monster" reads similar enough to the versions of the Egede UMO ("horrible sea-creature" and "dreadful monster") to give credence to Batchelder's supposition that the Spanish writer was familiar with, and indeed based his fictional account upon, the 1734 encounter, just as Batchelder himself did for his own 1849 book.

Afterword

A question remains concerning the precedents for, or the uniqueness of, the Egede UMO in the history of sea serpent lore. Oretelius' *Staukul* UMO, described and illustrated in his 1590 atlas (Figure 1.3), has already been mentioned as a possible precursor and inspiration. An even more striking couplet of related harbingers are described in the thirteenth-century document, the *King's Mirror*, which may have been known to the Egedes, and which is now heralded for its ability in providing accurate and rational descriptions of the fauna and natural phenomena of Iceland and Greenland (e.g., Nansen 1911; Whitaker 1986). Three anomalies—the anonymous writer uses the expression "marvels"—have, according to Lehn and Schroeder (2004:121), "defied identification and have thus been dismissed as legends bordering on the supernatural." Two of these are the *hafstramb*, or merman, and the *margygr*, or mermaid, both of which are to be found in Greenland waters where they display some traits in common to the Egede UMO:

> It is reported that the waters about Greenland are infested with monsters, though I do not believe that they have been seen very frequently. Still, people have stories to tell about them, so men must have seen or caught sight of them. It is reported that the monster called merman [*hafstrambr*] is found in the seas of Greenland. This monster is tall and of great size and rises straight out of the water. It appears to have shoulders, neck and head, eyes and mouth, and nose and chin like those of a human being; but above the eyes and the eyebrows it looks more like a man with a peaked helmet on his head. It has shoulders like a man's but no hands. Its body apparently grows narrower from the shoulders down, so that the lower down it has been observed, the more slender it has seemed to be. But no one has ever seen how the lower end is shaped, whether it terminates in a fin like a fish or is pointed like a pole. The form of this prodigy has, therefore, looked much like an icicle. No one has ever observed it closely enough to determine its body has scales like a fish or skin like a man. Whenever the monster has shown itself, men have always been sure that a storm would follow. They also noted how it turned when about to plunge into the waves ...

Another prodigy called mermaid [*margygr*] has also been seen there. This appears to have the form of a woman from the waist upward, for it has large nipples on its breast like a woman, long hands and heavy hair, and its neck and head are formed in every respect like those of a human being. The monster is said to have large hands and its fingers are not parted but bound together by a web like that which joins the toes of water fowls. Below the waist line it has the shape of a fish with scales and tail and fins. It is said to have this in common with the one mentioned before, that it rarely appears except before violent storms. Its behavior is often somewhat like this: it will plunge into the waves and will always reappear with a fish in its hands... The monster is described as having a large and terrifying face, a long sloping forehead and wide brows, a large mouth and wrinkled cheeks (Lehn and Schroeder 2004:121–122; Whitaker 1986:6–7).

Features of these two *King's Mirror* UMOs that are similar to the Egede UMO include: being large: (*King's Mirror* [KM]: "tall and of great size"; Egede [E]: "enormous big creature"), eye-hopping or breaching behavior (KM: "rises straight out of the water" and "turned when about to plunge into the waves"; E: "which rose itself high over the water" and "when it dived under it threw itself backward"), distinctive narrowing of the anterior body (KM: "shoulders, neck and head"; E: "head narrower than the body"), cuspated head (KM: "like ... a peaked helmet" and "long sloping forehead"; E: "long pointed nose"), general fusiform body shape (KM: "like an icicle"; E: the Bing illustration), prodigious fore-flippers (KM: "long hands ... bound together by a web"; E: "big broad flippers"), columnar posterior body shape (KM: "the lower down it has been observed, the more slender it has seemed to be"; E: "created at the rear like a serpent"), and shared body attributes of "wrinkles," "scales," "fins," and a long "tail," the latter speculated to be "pointed like a pole" (KM) or to be an observed "long tail" (E).

In one respect, the remarkable concordance in wording is such that it raises the possibility that the Egedes' descriptions might not have been as independently derived as has been commonly presumed. Alternatively, the concordance between the two sets of descriptions suggests that the Egede UMO may not be as unique as has hitherto been believed. In this respect, it is important to note that the descriptions of marine fauna in the *King's Mirror* correspond to typologies of generic animals that were frequently observed in Icelandic and Greenland waters; i.e., they were *not* isolated reports of solitary sightings, as is the case for the Egede UMO. It is a possibility that the characteristic narrowing terminus noted for the *hafstramb* in the *King's Mirror* may also represent a megafauna species, either cetacean or elasmobranch, observed pulling a train of entangled floats from hunting equipment or fishing gear. The opportunity was certainly there, as whales and sharks have both been hunted in Greenland and Iceland since at least the Middle Ages by Indigenous Inuit (e.g., Douglas et al. 2004; McCartney 1980; Taylor 1979; Whitridge 1999) and Nordic peoples (e.g., Gullov et al.

2010; MacNeil et al. 2012; Szabo 2008; Whitaker 1984) contemporaneous to fishing nets being set throughout the Northeastern Atlantic (e.g., Fagan 2017; Perdikaris and McGovern 2008, 2009; Seaver 1996).

We will, of course, never know the long-term fate of the Egedes' "very horrible sea-creature." But we can posit some suppositions. Unlike present-day animals, wherein entanglement in plastic material lasts for a lifetime, those in the eighteenth century are unlikely to have been so encumbered by trailing debris for much more than a single year, or at the most several years, due to the deterioration of natural fiber ropes and netting. But that being said, the experience would, just as for today, not have been a pleasant one, since such animals could have been disabled due to impaired swimming and feeding, and could have suffered the increased risk of infection from abrasion wounds (Derraik 2002). On the other hand, that the entangled whale "Necklace" is known to have returned to the Bay of Fundy for a number of years, and that the famous Gloucester Sea Serpent, which was almost certainly some entangled marine animal (Fama 2012; France 2019a), haunted its eponymous harbor for several years, suggests, if one is an optimist, that the Egede creature might have survived its limited period of entanglement to live out its natural existence. But again, due to the rapid expansion of whaling and shark hunting in the eighteenth century, this was a particularly precarious time to be such megafauna, holding as each and every individual did, such a considerable and easily exploitable resource of valued commodity and ensuing wealth in the form of body oil. Not until petroleum extraction began in 1859, following the oil strike in Pennsylvania, did these marvelous giants get a slight reprieve from the centuries of rapacious pursuit and plunder.

Sir Issac Newton famously (i.e., referenced in poetry by giants such as Byron and Milton, and in prose by countless of lessers ever since, such as in my own Ph.D. thesis) compared his scientific efforts to being "like a boy playing on the seashore, and diverting myself in now and then finding a smoother pebble or a prettier shell than ordinary, whilst the great ocean of truth lay all undiscovered before me." Regardless of what it might have actually been, the Egede UMO will continue to inspire the way in which we regard those undiscovered mysteries of the sea. For *in mari multa latent* goes the old adage, "in the ocean many things are hidden."

John Steinbeck (1968:27) was certainly correct in his contention of the need for such mystery in our modern consciousness: "There is some quality in man which makes him people the ocean with monsters and one wonders whether they are there or not. In one sense they are, for we continue to see them." Steinbeck next describes his excitement on hearing that "the decomposed body of a sea-serpent" had washed up on a nearby beach; but then admits his disappointment on learning that a reporter, who had rushed to the scene, found a note pinned to the decomposing carcass of "the evil-smelling monster" that read: "'Don't worry about it, it's a basking shark. [Signed] Dr. Rolph Bolin of the Hopkins Marine Station.'" Steinbeck continues:

> Dr. Bolin's kindness was a blow to the people of Monterey. They so wanted it to be a sea-serpent. Even we hoped it would be. When sometimes a true sea-serpent, complete and undecayed, is found and caught, a shout of triumph will go through the world. 'There, you see,' men will say, 'I knew they were there all the time. I just had a feeling they were there' (Steinbeck 1968:27).

The author concludes that

> Men really do need sea-monsters in their personal oceans ... For the ocean, deep and black in the depths, is like the low dark levels of our own minds in which dream symbols incubate and sometimes rise up ... And even if the symbol vision is horrible, it is there and it is ours. An ocean without its monsters would be like a completely dreamless sleep" (Steinbeck 1968:28–29).

Steinbeck's use of the words "horrible" and "sometimes rise up" of course call to mind Poul Egede's description of the UMO being "a very *horrible* sea-creature which *rose itself* so high over the water."

In the end, we owe much of our sense of the enduring mystery of the marine world, as romanticized by Newton, Steinbeck, and countless others, to that fateful encounter on July 6th, 1734, in the frigid waters off the remote coast of Greenland. And it is possible that entanglement may have played an important role in fostering that sense of the unknown by making what was seen to be imagined to be an unknown monster. Monsters are always there, swimming around in our collective imagination. For example, Philip Hoare (2010:225), commenting on the *Pauline* UMO—which I believe to have been, as for the Egede creature, an encounter with an entangled animal (France 2016c)—echoes Steinbeck's sentiments:

> I must confess I have seen whales that look like sea monsters, rolling in the waves. My childhood desire to believe in a lost world (Arthur Conan Doyle, on honeymoon in Greece, claimed to have seen a young ichthyosaur in the sea) seeks to create something palpable out of the apparently incredible; to conjure an abyssal nightmare out of the pages of scientific certainty.

Therefore, although they may not be corporeal entities, sea monsters remain very much "real" in the sense of their conceptual role in influencing social history.

Appendix 1.

Here [Again] *are Dragons*: Locations for Further Ethnozoological Explorations

As correctly contended, in voices as varied as novelist John Steinbeck, science writer Philip Hoare, and folklorist Peter Dendle, sea monsters in modern times satisfy a deep-rooted longing in the human psyche to believe in oceanic mysteries. Consequently, cryptozoological books about sea monsters continue to be published at a rapid pace, added to which are more than a two dozen thriller-type novels, all coming out within the last several years.

In their *The Field Guide to Lake Monsters, Sea Serpents, and Other Mystery Denizens of the Deep* (2003), cryptozoologists Coleman and Huyghe (2003) conclude with an Afterword entitled "Top Eight Places to Look for a Sea Serpent," wherein they offer "a few random suggestions" as to where to find cryptids, believing as they do in the actual existence of such animals. However, they caution that because such beings are by definition very elusive, "to find a good place to view a Sea Serpent is not an easy task" (Coleman and Huyghe 2003:289). Locations described include Cadboro Bay, Vancouver Island (believed home of "Caddy"), Soay Island in the Orkneys (where the "Soay Beast" was spotted), Chesapeake Bay (residence of "Chessie"), and the north Massachusetts coast, near Cape Ann (home of the "Gloucester Sea Serpent"). Given their enduring allure, the following suggestions, including photographic evidence, are offered as locations where interested ethnozoologists can be guaranteed to observe sea monsters, some artful constructs, others, very much authentic entities; in short the "Top Eight Places to Look for a [*Real*] Sea Serpent."

1. Natural History Museum, London, England

It is but a short walk from the Tube station through South Kensington to stand in front of the ornate architecture of the Victorian Era's most notable "cathedral of science," the Natural History Museum. This is "sea-serpent slayer" Sir Richard Owen's greatest legacy: an open access museum to educate the masses about science. Inside, one moves past the statue of Darwin looming down into the main hall from his staircase perch. Ignoring the famous *Diplodocus* replica skeleton, the suspended life-size whale models, and the animatronic T-rex, the latter surrounded by a gaggle of captivated children, one arrives in the Fossil Marine Reptile gallery. Here, mounted on a long wall is the largest concentration of fossils (many being casts of the originals) found by Mary Anning in the cliffs of Lyme Regis. These are some of the most complete fossils of ichthyosaurs, pliosaurs, and plesiosaurs in existence. It is fascinating to examine details of their teeth, eye discs, and fins (Figure A.1). And it is humbling to be here, where the discipline of paleontology began, and view the fossils that inspired so many nineteenth-century sightings of "sea dragons," which in turn inspired the "prehistoric fiction" by the likes of Verne, Conan Doyle, and Rice Burroughs, that went on to inspire the famous, and now-acknowledged faked "surgeon's photo" of the Loch Ness Monster, all of which together have fueled generations of cryptozoological hope and histrionics.

Figure A.1. Casts of marine fossils in London's Natural History museum.

2. Harvard Museum Natural History, Cambridge, Massachusetts

Harvard University's Museum of Natural History arose in 1998 as the public face of three amalgamated research museums, each dating back to the nineteenth century. Although it is tempting to be sidetracked by the internationally renowned glass flowers (teaching models whose exquisite details absolutely beggar belief), those seeking sea monsters need to proceed to the Romer Hall of Vertebrate Paleontology. There, behind one of the first *Triceratops* skulls ever discovered, is the world's largest turtle shell, an automobile-sized fossil from the prehistoric freshwater species *Stupendemys geogrophicus.* Knowing there to have been a long tradition of mistaking large sea turtles, such as leatherbacks, for sea monsters, once can savor viewing and touching the fossil. But the real reason that all sea monster aficionados should visit is to be found on the other side of the room. For there, running for 14 m along the complete length of a wall, is the world's only mounted *Kronosaurus*, found during a 1931 expedition to Australia (Figure A.2). Although recent research has suggested that too many vertebrae were included in the reconstruction, resulting in its length shrinking to about 10 m, there is still no denying that this is one truly fearsome beast. Mostly, this results from the impressive—and scary looking!—"business end": the 3 m long skull, filled as it is with enormous teeth. This monstrous short-necked pliosaur from the Cretaceous Period was one of the all-time greatest apex predators, making Mary Anning's more diminutive plesiosaurs and ichthyosaurs seen in London, seem like unthreatening pets by comparison.

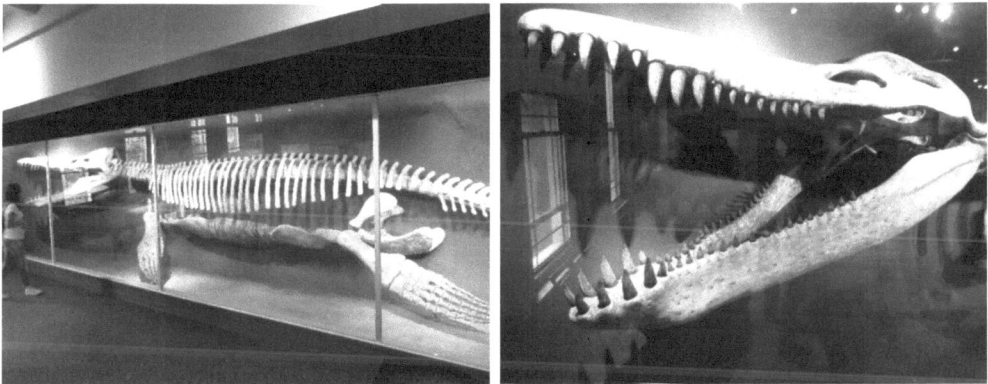

Figure A.2. *Kronosaurus* fossil at Harvard University's Museum of Natural History.

3. Canadian Fossil Discovery Centre, Morden, Manitoba

Despite Manitoba having a subarctic coastline of 600 km length, because it is so distant and difficult to access for the majority of the province's population that live close to the American border, the ocean seems such a remote and totally foreign place. But go back 80 million years, during the Cretaceous, and the flat prairie provinces would have been submerged beneath a massive sea that occupied most of what is now the central part of the North American continent. Visiting the town of Morden, located in the middle of wheat fields south of Winnipeg, likewise seems incongruous to the present search for sea monsters. But the museum there contains the largest collection of marine fossils in Canada. These include an 8 m long fossil of a pliosaur, and the 2 m long fossil of the much rarer *Hesperornis*, a giant flightless and toothed marine bird. But the reason why so many go to visit the Fossil Discovery Centre, is to see "Bruce" and "Suzy," which are, respectively, the world's largest, at 14 m, and the world's most complete, fossils of mosasaurs (Figure A.3). Sporting shark-like teeth and a tail fin, the elongated and fusiform mosasaur is, as the museum boasts, a "huge, scary, flesh-eating lizard often called the T. rex of the sea." Of all prehistoric marine beasts, it is these that most easily assume the title of being "sea monsters" par excellence. For any who find it difficult to conceptually imagine what a living mosasaur would look like from staring at the mounted array

of reassembled fossil bones in the museum, a nearby bus-sized replica model of Bruce (Figure A.3), guarantees to be the stuff of sea monster nightmares. Bruce is so named as, at its time of discovery, the fossil hunters were enamored by the famous Monty Python sketch wherein everyone has that moniker.

Figure A.3. Fossil of a pilosaur and fossil replica model of "Bruce" at the Canadian Fossil Discovery Centre.

4. Archeological Museum, Thessaloniki, Greece

Hellenic and Hellenistic Greeks, if not the inventors of sea serpents in the Western World—that honor goes to the ancient Mesopotamians—were certainly, in all of antiquity, those whom most populated their "wine-dark seas" with a myriad of aquatic gods and enormous monsters. It is no accident that whales belong to the order Cetacea, derived from the ancient Greek word *ketos,* meaning sea monster. In consequence, most Greek museums contain pottery, sculpture, and mosaics that depict all manner of strange and large marine fauna, enough to set any sea serpentologist's or cryptozoologist's heart aflutter. The museum in Thessaloniki, for example, has the remains of an impressive wall fresco that portrays a fisherman surrounded by a bountiful sea full of creatures, several of which are much larger than his boat (Figure A.4). These mysterious animals are drawn as sinuous, elongated *ketos* with two sets of small fins, resembling as they do, giant prehistoric mososaurs. Then there is the funerary stela which portrays two robed Greeks confronting a massive coiled serpent sporting a billy-goat beard, and resembling, therefore, a number of nineteenth-century representations of sea serpents. In the museum shop, one can purchase, in addition to a replica stela of this giant serpent (Figure A.4), the book *The Sea of Gods, Heroes and Men in Ancient Greek Art* (Samara-Kauffmann 2008). The chapter "The Sons of Poseidon and Other Sea Monsters" contains examples of decorative art incorporating sea monsters. One vase depicts the sea demon Triton looking like an anthropomorphized version of the stela serpent. And another shows Herakles or Perseus fighting a ferocious sea monster with a fish-like undulating body and a wolf head (Figure A.4). Also shown, thereby giving a sense of verisimilitude to the scene, are recognizable marine animals such as dolphins, and an octopus and seal.

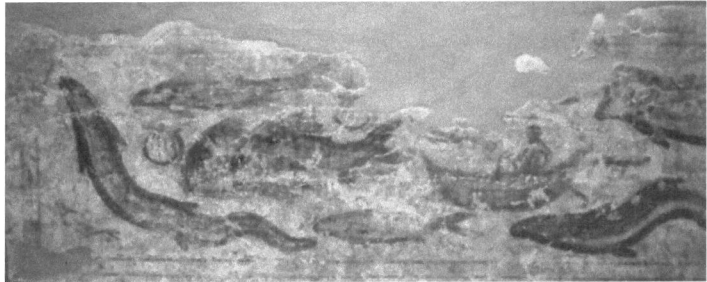

Figure A.4. Depictions of mysterious sea creatures at the Thessaloniki Archeological Museum.

5. Crystal Palace, Sydenham, London, England

To stand in "Dinosaur Park" at this location is to be at the very nexus of the public's enduring fascination with prehistoric creatures. Here one is able to appreciate how, with little difficulty, this site was responsible for giving birth to the Victorian idea that certain marine creatures might have somehow escaped extinction, persisting into the modern age as sea monsters. Named after the cast iron and glass building that housed the Great Exhibition (today referred to as the World's Fair) of 1851, which had been moved to this newly developed location in the suburbs, the park became a popular Victorian pleasure ground. Here, from 1854 onwards, Sunday afternoon strollers were able to view life-models of 15 genera of extinct animals that had been designed and sculpted by Benjamin Waterhouse-Hawkins under the careful supervision of paleontologist Sir Richard Owen. Today, the park is open every day and is filled with milling families pointing fingers at, and reading interpretative signs about, the various creatures displayed on a series of chronologically-themed islands situated in a large pond. To the knowledgeable observer, the models bring a smile, inspired as they were from scant bones and much imagination, as for example, the spikey thumb of the *Iguanodon* being misplaced as a horn atop its nose. The marine reptiles are also inaccurately represented. Ichthyosaurs with odd-looking tails bask on a beach like lazy sea lions, and plesiosaurs twist their necks into impossible contortions while lulling in the shallows (Figure A.5). But again, these are the world's first models of such extinct animals, created a decade before Darwin's *magnum*

opus. For this is a place to celebrate imagination, given that more than a dozen works of fiction have been produced in which the models come to life. Living, non-extinct prehistoric creatures: the dream of many a cryptozoologist. What could be better than that?

Figure A.5. Crystal Palace "dinosaur" models from the Great Exposition of 1851.

6. Wookey Hole, Somerset, England

The contrast between the Crystal Palace of yesteryear and the Wookey Hole of today provides all the evidence needed to satisfy those pessimists who believe that modern civilization is in rapid decline. The site contains a series of impressive limestone caverns in the Mendip Hills that have been occupied by humans since the Paleolithic. A medieval legend of a nasty witch turned into a stalagmite by a monk has imbued the area with a fairytale atmosphere that, since 1927, has been capitalized through the creation of a Disneyesque tourist monstrosity that puts to bed the prejudiced notion that such tackiness is restricted to and celebrated in any single, particular country. Past the hotel, the hordes of children, the penny arcades, the house of mirrors, the chamber of horrors, the pirate-themed mini-golf course, and such, the determined sea serpent seeker will finally see signage for the "Valley of the Dinosaurs." And there, admix the dated and gaudy fiberglass models of life-sized dinosaurs—all the expected usual suspects; i.e., *Brontosaurus, Pteranodon, Tyrannosaurus*, etc., (the latter labelled as "the terrible child eating T. Rex" no less)—are a set of models of prehistoric marine reptiles that have been tucked into the background foliage (Figure A.6). Instead of being disappointed, for the knowledgeable sea serpent aficionado, such dino-kitsch seems appropriate given the touristic hucksterism that often played such an important part in the cultural history of many sea serpent and lake monster phenomena.

Figure A.6. Contemporary models of prehistoric marine reptiles at Wookey Hole.

7. Bayerisches Staatsbibliothek, Munich, Germany

Created in Rome by exiled Swedish Catholic priest Olaus Magnus, and published in Venice in 1539, the *Carta marina et descriptio septentrionalium terrarium* ("Marine map and description of the Northern lands"), *Carta marina* for short, was the largest and most detailed map for any region of Europe produced up to that time (Figure A.7). At nearly 2 m wide, and composed of nine black and white folio sheets from woodcut blocks, only two copies are known to exist. One of these was found in 1886 and is housed in the national library in Munich. Another 1572 colored version is easily found online and is used as the foundation in Joseph Nigg's wonderful 2013 book *Sea Monsters: A Voyage Around the World's Most Beguiling Map*. And so it is to Munich that one must travel to go to the library in a hope to be given access to see the original, for there is no more important source demonstrating the European fascination with sea monsters than those nine sheets of paper. Unfortunately, even scholars are unable to see the original, instead having to be satisfied with viewing life-sized facsimile panels printed onto heavy paper stock. And even then, in order to see what are, after all, just copies, one must fill out numerous forms in order to satisfy what the novelist Phillip Kerr refers to as the German predilection for documentation archiving. Once the laborsome "administriva" is completed, the dedicated sea serpent scholar is allowed to carefully examine the facsimile panels at a table situated proximal to the omnipresent surveillance of the supervising archivist. After first recognizing the names and locations of known (and possibly visited) cities, one's eyes are drawn to the parts of the map showing offshore waters, as it is here where almost two dozen sea monsters are drawn (Figure A.7). So on a map that portrays the most accurate rendering of Nordic cartography made up until that time (and for many decades thereafter), are also displayed a menagerie of terrifying marine creatures. Cryptozoologists love this sort of thing, for if the former is deemed correct and believable, why not then the latter? Magnus obtained his information from fishermen and sailors and accompanied the map, in 1555, with a Latin text describing the Northern peoples and their environment. In this, he writes: "For in the Ocean that is so broad and by an easie and fruitful increase receives the Seeds of Generation, there are found many monstrous things in sublime Nature." In particular, the eyes of the sea serpent seeker linger over the representations of the "Sea Orm" and the sea serpent (Figure A.7), both later to be mentioned by Bishop Pontoppidan in his seminal work.

Figure A.7. The *Carta marina*, (ir. Nigg 2013:13) a copy of which exists at the Bayerisches Staatsbibliothek.

8. One's Own Living Room

But the best place to observe sea monsters might very well be in the comforts of one's own living room. From there, one can download and watch the 2003 documentary, and then read the accompanying 2004 book, *Chased by Sea Monsters: Prehistoric Predators of the Deep*, both produced by the BBC. Created by Nigel Marven and Jasper James, those behind the widely popular *Walking with Dinosaurs* series, the film and book bring to life, in a humorous yet educational way, what it would have been like to cohabit with the monsters swimming about the World's "deadly prehistoric seas." Transported on a time-travelling sailing ship, our intrepid hero physically immerses himself—sometimes free swimming, but often within the well-needed security of a shark cage—in order to record encounters with the denizens of "Hell's Aquarium" in a countdown as to which particular epoch's apex predators would have been the all-time nastiest. Spectacular filmed computer graphics, captured as wonderful stills in the coffee-table book (Figure A.8), provide a fascinating portrayal of what it would have been like for a measly human to confront such behemoths. Supported by the latest scientific evidence, this combination of film and page is the modern descendent of more than a century of prehistoric fiction (*Pellucidar* never looked so frighteningly uninviting). And if that isn't frightful enough, one can watch the 2012 documentary *Titanoboa: Monster Snake*, which includes scary computer graphics of a prehistoric aquatic snake that approaches the same proportions as those ascribed to contemporary sea serpents by cryptozoologists. Once finished viewing these efforts, it will be a brave person indeed who has the resolve to venture out into coastal waters on a small boat … for perhaps there is just a remote chance that there really *is* something out there after all ….

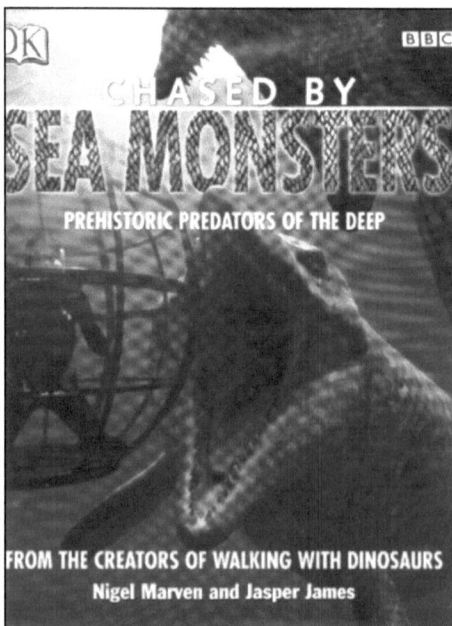

Figure A.8. Book *Chased by Sea Monsters: Prehistoric Predators of the Deep* (Marven and James 2004).

References Cited

Adamnan. 1856. *The Life of Saint Columba*. University Press for the Irish Archaeological and Celtic Society, Dublin, Ireland.

Aiken, W.R.G., and J. Purser. 1936. The Preservation of Fibre Ropes for Use in Sea-Water. *Plymouth Laboratory New Series* 20:643–654.

Al-Abdulrazzak, D., R. Naidoo, M.L.D. Palomares, and D. Pauly. 2012. Gaining Perspective on What We've Lost: The Reliability of Encoded Anecdotes in Historical Ecology. *PLOS One* 7:1–5.

Alexander, K.E., W.B. Leavenworth, T.V. Willis, C. Hall, S. Mattocks, S.M. Bittner, E. Klien, M. Staudinger, A. Bryan, J. Rosset, B.H. Carr, and A. Jordan. 2017. Tambora and the Mackerel Year: Phenology and Fisheries During an Extreme Climate Event. *Science Advances* 3:1–18.

Americana Media Specials. 2019. *Mysteries of the Loch Ness Monster*. Americana Media Specials, New York, NY.

Anderson, E.N., D. Pearshall, E. Hunn, and N. Turner, eds. 2011. *Ethnobiology*. Wiley-Blackwell, Hoboken, NJ.

Angenot, M., and N. Khouri. 1981. An International Bibliography of Prehistoric Fiction. *Science Fiction Studies* 8:38–53.

Arment, C. 2004. *Cryptozoology: Science & Speculation*. Coachwhip Publications, Darke County, IA.

Ashton, J. 1890. *Curious Creatures in Zoology*. John C. Nimmo Publishers, London, UK.

Asma, S.T. 2009. *On Monsters: An Unnatural History of Our Worst Fears*. Oxford University Press, Oxford, UK.

Australian, The. 2019. Rare 3500 kg Basking Shark Caught is Donated to Science. *The Australian*. 23 June:4. Sydney, Australia.

Barber, L. 1980. *The Heyday of Natural History 1820–1870*. Doubleday & Company, New York, NY.

Barclay, J. 1811. Remarks on Some Parts of the Animal that Was Cast Ashore on the Island of Stronsa. *Memoirs of the Wernerian Natural History Society* 1:418–444.

Bartholomew, G.A. 1986. The Role of Natural History in Contemporary Biology. *BioScience* 36:324–329.

Bartholomew, R.E. 2012. *The Untold Story of Champ: A Social History of America's Loch Ness Monster*. Excelsior Editions, Albany, NY.

Batchelder, E. 1849. *Romance of the Sea Serpent, or, the Icthyosaurus*. Barlett, Burlington, MA.

BBC News. 2019. Whale Washed Up in Caithness Tangled in Canadian Fishing Gear [web page]. URL: https://www.bbc.com/news/uk-scotland-highlands-islands-48497046. Accessed on January 20, 2019.

Beard, D. 1887. *Sea Monsters*. Harper and Brothers, New York, NY.

Berger, C. 1983. *Science, God, and Nature in Victorian Canada*. University of Toronto Press, Toronto, ON, Canada.

Berger, M.J. 2000. *Thoreau's Late Career and 'The Dispersion of Seeds': The Saunterer's Synoptic Vision*. Camden House, Rochester, NY.

Berkowitz, C., and B. Lightman, eds. 2017. *Science Museums in Transition: Cultures of Display in Nineteenth-Century Britain and America*. University of Pittsburgh Press, Pittsburgh, PN.

Berlin, B. 1992. *Ethnobiological Classification: Principles of Categorization of Plants and Animals in Traditional Societies*. Princeton University Press, Princeton, NJ.

Bigelow, J. 1820. Documents and Remarks Respecting the Sea-Serpent. *American Journal Science and Arts* II:147–154

Boas, F. 1904. The Folk-lore of the Eskimo. *Journal of American Folklore* 17:1–13.

Bolster, W. J. 2012. *The Mortal Sea: Fishing the Atlantic in the Age of Sail*. Harvard University Press, Cambridge, MA.

Boskovic, R., K.M. Kovacs, M.O. Hamilton, and B.N. White. 1996. Geographical Distribution of Mitochondrial DNA Haplotypes in Grey Seals (*Halichoerus grypus*). *Canadian Journal of Zoology* 74:1787–1796.

Bright, M. 1989. *There are Giants in the Sea*. Robson, London, UK.

Brink-Roby, H. 2008. Siren Canora: The Mermaid and the Mythical in Late Nineteenth-Century Science. *Archives of Natural History* 35:1–14.

Brito, C., N. Viera, and J.G. Freitas. 2019. The Wonder Whale: A Commodity, a Monster, a Show and an Icon. *Anthopozoologica* 54:13–27.

Brongersma, L.D. 1968. The Soay Beast. *Beaufortia* 15:33–46.

Brown, A. 2019. Dead Whale Was Tangled in Rope in East Lothian for 'Months' [web page]. BBC News 25 April. URL: https://www.bbc.com/news/uk-scotland-edinburgh-east-fife-48051954. Accessed January 20, 2019.

Brown, C.M. 1990. A Natural History of the Gloucester Sea Serpent: Knowledge, Power, and the Culture of Science in Antebellum America. *American Quarterly* 42:402–436.

Brown, R. 1868. On the Mammalian Fauna of Greenland. *Proceedings of the Zoological Society of London* (May):330–362.

Browne, J.R. 1846. *Etchings of a Whaling Cruise*. Harper and Brothers Publishing, New York, NY.

Burnett, D.G. 2007. *Trying Leviathan: The Nineteenth-Century Court Case that Put the Whale on Trial and Challenged the Order of Nature*. Princeton University Press, Princeton, NJ.

Burns, E.I. 2014. Monster on the Margin: The Sea Serpent Phenomenon in New England, 1817–1849. Unpublished Doctoral Dissertation, Department of History, University of Buffalo, Buffalo, NY.

Burton. M. 1954. *Living Fossils.* Thames & Hudson, London, UK.

Butcher, D. 2000. The Herring Fisheries in the Early Modern Period: Lowestoft as Microcosm. In *England's Sea Fisheries: The Commercial Sea Fisheries of England and Wales Since 1300,* edited by D.L. Starkey, J. Ramster, and C. Reid, pp. 54–63, Chatham Publishing, Newbury, UK.

Carrington, R. 1957. *Mermaids and Mastodons: A Book of Natural and Unnatural History.* Chatto and Windus Publishers, London, UK.

Cazeils, N. 2011. *Musée du Fjord Sea Monsters.* Les Editions GID, Montreal, QC, Canada.

Cheezum, E.A. 2007. Discovering Chessie: Waterfront, Regional Identity, and the Chesapeake Bay Sea Monster, 1960–2000. Unpublished Doctoral Dissertation, Department of History, University of South Carolina, Columbia, SC.

Coleman, L., and P. Huyghe. 2003. *The Field Guide to Lake Monsters, Sea Serpents, and Other Mystery Denizens of the Deep.* Putman, New York, NY.

Conybeare, W.C. 1824. On the Discovery of an Almost Perfect Skeleton of *Plesiosaurus. Transactions of the Geological Society of London.* 2:382–389.

Cope, E.D. 1869a. Synopsis of the Extinct Batrachia, Reptilia and Aves of North America. *Transactions of the American Philosophical Society.* XIV:1–122.

Cope, E.D. 1869b. The Fossil Reptiles of New Jersey. *American Naturalist* 3:84–91.

Costello, P. 1974. *In Search of Lake Monsters.* Coward, McCann & Geoghegan, New York, NY.

Cotton, P.A., D.W. Sims, S. Fanshawe, and M. Chadwick. 2005. The Effects of Climate Variability on Zooplankton and Basking Shark (*Cetorhinus maximus*) Relative Abundance off Southwest Britain. *Fisheries Oceanography* 14:151–155.

Crafts, W. 1819. *The Sea Serpent, or Gloucester Hoax: A Dramatic Jeu d'Esprit in Three Acts.* A.E. Miller, Charleston, SC.

Crantz, D. 1767. *The History of Greenland.* Cambridge University Press, Cambridge, UK.

Crumley, C.J., ed. 1994. *Historical Ecology: Cultural Knowledge and Changing Landscapes.* School for Advanced Research Press, Santa Fe, NM.

Das, N. 2009. Occam's Razor and Cryptozoology. *The Scientia Review.* Massachusetts Institute of Technology.

da Silva Vieira, K., W.L. da Silva Viera, and R.R. Alves. 2017. Imaginary Zoology: Mysterious Fauna in the Reports of Ancient Travelers and Chroniclers. In *Ethnozoology: Animals in Our Lives,* edited by R.T. Alves. and U.P. Albuquerque, pp. 23–56. Elsevier, London, UK.

Davidson, J.P. 2002. Bonehead Mistakes: The Background in Scientific Literature and Illustrations for Edward Drinker Cope's First Restoration of *Elasmosaurus platyurus. Proceedings of the Academy of Natural Sciences of Philadelphia* 152:215–240.

Davies, S. 2016. *Renaissance Ethnography and the Invention of the Human: New Worlds, Maps and Monsters*. Cambridge University Press, Cambridge, UK.

Davis, V.A. 1888. Serpent Myths. *The North American Review* 375:161–171.

de Camp, L.S., and C.C. de Camp. 1985. *The Day of the Dinosaur*. Bonanza Books, New York, NY.

Deedy, A. 2017. Hark! A Sea Monster! (Oh, No, Just a Dying Whale). All Those Strange Sea Monster Sightings in Days of Yore? This May Be the Best Explanation Yet. *Hakai Magazine*. [online] URL: https://www.hakaimagazine.com/news/hark-sea-monster-oh-no-just-dying-whale/. Accessed on January 22, 2019.

Del-Claro, K., V. Stefani, D. Lange, A.A Vilela, L. Nahas, M. Velasques, and H.M. Torezan-Silingardi. 2013. The Importance of Natural History Studies for a Better Comprehension of Animal-Plant Interaction Networks. *Bioscience Journal* 29(2). DOI:10.14393/BJ-v29n2a2013-17892.

Dendle, P. 2006. Cryptozoology in the Medieval and Modern Worlds. *Folklore* 117:190–206.

Derraik, J.G.B. 2002. The Pollution of the Marine Environment by Plastic Debris: A Review. *Marine Pollution Bulletin* 42:842–852.

Dewdney, S., and K.E. Kidd. 1973. *Indian Rock Paintings of the Great Lakes*. University of Toronto Press, Toronto, ON, Canada.

Dixon, D.P., and S.M. Ruddick. 2013. Monsters, Monstrousness, and Monstrous Nature/s. *Geoforum* 48:237–238.

Douglas, M., J.P. Smol, J.M. Savelle, and J.M. Blais. 2004. Prehistoric Inuit Whalers Affected Arctic Freshwater Ecosystems. *Proceedings of the National Academy of Science* 101:1613–1617.

Drinnon, D. 2010. Egede Sea Serpent Sighting. *Still on the Track: The Voice of the International Cryptozoological Community*. [online] URL: https://forteanzoology.blogspot.com/2010/04/dale-drinnon-egede-sea-serpent-sighting.html. Accessed on January 23, 2019.

Duhamel du Monceau, H-L. 1769. *Traité Général des Pêches*. La Marre, Paris, France.

Eberhart, G.M. 2002. *Mysterious Creatures: A Guide to Cryptozoology*. ABC-CLIO Press, Santa Barbara, CA.

Egede, H. 1741. *Grønlands nye Perlustration eller Naturel Historie*. Johan Christoph Groth, Copenhagen, Denmark.

Egede, H. 1745. *A Description of Greenland*. C. Hitch, London, UK.

Egede, P. 1741. *Continuation af den Grønlandske Mission. Forfattet I form a fen Journal fra Anno 1734 til 1740*. Johan Christoph Groth, Copenhagen, Denmark.

Egede, P. 1789. *Efterretninger om Grøonland uddragne af en Journal fra 1721 til 1788*. Hans Christopher Schrøoder, Copenhagen, Denmark.

Eggleton, B., and N. Suckling. 1998. *The Book of Sea Monsters*. The Overlook Press, New York, NY.

Ellis, R. 1994. *Monsters of the Sea.* Alfred A. Knopf, New York, NY.

Ellis, R. 2003. *Sea Dragons: Predators of the Prehistoric Oceans.* University Press of Kansas, Lawrence, KS.

Elman, R. 1977. *First in the Field: America's Pioneering Naturalists.* Van Nostrand Reinhold, New York, NY.

Elwen, S., and T. Gridley. 2013. Gray Whale (*Eschrichtius robustus*) Sighting in Namibia (SE Atlantic)—First Record for Southern Hemisphere. *Whaling Commission Scientific Committee* SC/65a.

Emling, S. 2009. *The Fossil Hunter: Dinosaurs, Evolution, and the Woman Whose Discoveries Changed the World.* Palgrave Macmillian, London, UK.

Enterline, J.R. 2002. *Erikson, Eskimos & Columbus: Medieval European Knowledge of America.* John Hopkins University Press, Baltimore, MD.

Evans, H.E. 1993. *Pioneer Naturalists.* Henry Holt & Company, New York, NY.

Fabricius, O. 1780 *Fauna Grøenlandica.* Göttingen Niedersächsisch, Staats.

Fagan, B. 2000. *The Little Ice Age: How Climate Made History 1300-1850.* Basic Books, New York, NY.

Fagan, B. 2017. *Fishing: How the Sea Fed Civilization.* Yale University Press, New Haven, CT.

Fairfax, D. 1998. *The Basking Shark in Scotland—Natural History, Fishery and Conservation.* Tuckwell Press, Glasgow, UK.

Fama, E. 2012. Debunking a Great New England Sea Serpent [online]. URL: https://www.tor.com/2012/08/16/debunking-a-great-new-england-sea-serpent/. Accessed on June 3, 2016.

Ferguson, M.A., R.G. Williamson, and F. Messier. 1998. Inuit Knowledge of Long-Term Changes in a Population of Arctic Tundra Caribou. *Arctic* 51:201–219.

Forth, G. 2016. *Why the Porcupine is Not a Bird: Explorations in the Folk Zoology of an Eastern Indonesian People.* University of Toronto Press, Toronto, ON, Canada.

Forth, G. 2020. Classifying Mermaids: Observations on Local Naming and Classification of Dugongs (*Dugong dugon*) among the Lio of Flores Island (Eastern Indonesia). *Journal of Ethnobiology* 40:56–69.

France, R.L. 2010. *High Arctic Extreme Science: Environmental Research from the Trans-Ellesmere Island Ski Expedition.* Green Frigate Books (Libri), Faringdon, UK.

France, R.L. 2016a. From Land to Sea: Governance-Management Insights from Terrestrial Restoration Research Useful for Developing and Expanding Social-Ecological Marine Restoration. *Ocean & Coastal Management* 133:64–71.

France, R.L. 2016b. Historicity of Sea Turtles Misidentified as Sea Monsters: A Case for the Early Entanglement of Marine Chelonians in Pre-Plastic Fishing Nets and Maritime Debris. *Coriolis: An International Journal of Maritime Studies* 6:1–24.

France, R.L. 2016c. Reinterpreting Nineteenth-Century Accounts of Whales Battling 'Sea Serpents' as an Illation of Early Entanglement in Pre-Plastic Fishing Gear or Maritime Debris. *International Journal of Maritime History* 28:686–714.

France, R.L. 2017. Imaginary Sea Monsters and Real Environmental Threats: Reconsidering the Famous Osborne, 'Moha-moha', Valhalla, and 'Soay Beast' Sightings of Unidentified Marine Objects. *International Review of Environmental History* 3:63–100.

France, R.L. 2018. Illustration of an 1857 'Sea-Serpent' Sighting Re-Interpreted as an Early Depiction of Cetacean Entanglement in Maritime Debris. *Archives of Natural History* 45:111–117.

France, R.L. 2019a. *Disentangled: Ethnozoology and Environmental Explanation of the Gloucester Sea Serpent*. Wageningen Academic Publishers, Wageningen, The Netherlands.

France, R.L. 2019b. Extreme Climatic Upheaval, Emergency Resource Adaptation, and the Emergence of Folkloric Belief: Geomythic Origin of Sea Serpents from Animals Becoming Entangled in Fishing Gear during New England's Nineteenth-Century Social-Ecological Crisis. *Human Ecology* 47:499–513.

France, R.L. 2019c. Ethnobiology and Shifting Baselines: An Example Reinterpreting the British Isles' Most Detailed Account of a Sea Serpent Sighting as Early Evidence for Pre-Plastic Entanglement of Basking Sharks. *Ethnobiology and Conservation* 8:1–31.

France, R.L. 2020a. From Cryptozoology to Conservation Biology: An Earlier Baseline for Entanglement of Marine Fauna in the Western Pacific Revealed from Historic 'Sea Serpent' Sightings. *Advances in Historical Studies* 9:45–69.

France, R.L. 2020b. Early Entanglement of Nova Scotian Marine Animals in Pre-Plastic Fishing Gear or Maritime Debris: Indirect Evidence from Historic 'Sea Serpent' Sightings. *Proceedings of the Nova Scotian Institute of Science* 50:319–349.

France, R., and M. Sharp. 1992. First Record of the Rough-Legged Hawk, *Buteo lagopus*, from Ellesmere Island, Northwest Territories. *Canadian Field Naturalist* 106:511–512.

France, R.L., and M. Sharp. 1995. Polynyas as Centers of Organization for Structuring the Integrity of Arctic Marine Communities. In *To Preserve Biodiversity – An Overview: Readings in Conservation Biology,* edited by D. Ehrenfeld, pp. 245–249. Blackwell Science, Cambridge, MA. Originally published 1992, *Conservation Biology* 6:442–446.

Franklin, A. 1990. *Cherokee Folk Zoology*. Garland Science, New York, NY.

Fuller, J. 2001. Before the Hills in Order Stood: The Beginning of the Geology of Deep Time in England. *Geological Society London Special Publications* 190:15–23.

Galbreath, G.J. 2015. The 1848 'Enormous Serpent' of the *Daedalus* Identified. *The Skeptical Inquirer* 39(5):42–46.

Gardner, D. 2007. Whale Survives Harpoon Attack 130 Years Ago to Become 'World's Oldest Mammal'. *The Daily Mail*, 13 June. London, UK.

Garnett, E. 1968. *To Greenland's Icy Mountains and the Story of Hans Egede, Explorer, Colonizer, Missionary*. Heinemann, Portsmouth, NH.

Gatschet, A.S. 1899. Water-Monsters of American Aborigines. *The Journal of American Folklore* 47:255–260.

Gessner, C. 1551–1558. *Historia animalium.* University of Zurich, Zurich, Switzerland.

Gibson, J. 1887. *Monsters of the Sea: Legendary and Authentic.* T. Nelson, Edinburgh, UK.

Gifford, G.E. 1965. Twelve Letters from Jeffries Wyman, M.D.: Hampton-Sydney Medical College, Richmond, Virginia, 1843–1848. *Journal of the History of Medicine and Allied Sciences* 20:320–322.

Gilchrist, G., M. Mallory, and F. Merkel. 2005. Can Local Ecological Knowledge Contribute to Wildlife Management? Case Studies of Migratory Birds. *Ecology and Society* 10(1). [online] URL: https://www.jstor.org/stable/26267752. Accessed on January 23, 2019.

Gore, M., L. Abels, S. Wasik, L. Saddler, and R. Ormond. 2019. Are Close-Following and Breaching Behaviors by Basking Sharks at Aggregation Sites Related to Courtship? *Journal of the Marine Biological Association of the United Kingdom* 99:681–693.

Gosse, P.H. 1860. *Romance of Natural History.* Sheldon and Company, New York, NY.

Gould, C. 1886. *Mythical Monsters.* W.H. Allen, London, UK.

Gould, R.T. 1930. *The Case for the Sea-Serpent.* Philip Allen, London, UK.

Green, J. 1743. A Description of Old and New Greenland, or a Natural History of Old Greenland's Situation, Air, Habitude, and Circumstances. *Royal Society Philosophical Transactions* 42:607–615.

Greener, M. 2010. The Golden Age of Sea Serpents. *Fortean Times* 3:34–39.

Gregory, M.R. 2009. Environmental Implications of Plastic Debris in Marine Settings – Entanglement, Ingestion, Smothering, Hangers-On, Hitch-Hiking and Alien Invasions. *Philosophical Transactions of the Royal Society B* 364:2013–2025.

Guidetti, P., and F. Micheli. 2011. Ancient Art Serving Marine Conservation. *Frontiers in Ecology and Environment.* 9:374–375.

Gulløv, H.C., J. Bjarke, T. Pedersen, B.H. Jakobsen, and A. Kroon. 2010. Commercial Hunting Activities in the Greenland Sea: The Impact of the European Blubber Industry on East Greenland Inuit Societies/Optically Stimulated Luminescence Dating of Inuit Settlement Structures in Coastal Landscapes of Northeast Greenland. *Geografisk Tidsskrift-Danish Journal of Geography* 110:357–371.

Hackett, J., and S. Harrington, eds. 2018. *Beasts of the Deep: Sea Creatures and Popular Culture.* John Libbey Publishing, Bloomington, IN.

Hamilton, R. 1839. *The Natural History of the Amphibious Carnivora: Including the Walrus and Seals, and the Herbivorous Cetacea, Mermaids, etc.* Jardine's The Naturalists Library. W.H. Lizars, London, UK.

Harrison, P. 2001. *Sea Serpents and Lake Monsters of the British Isles.* Robert Hale, London, UK.

Hawkins, T. 1840. *The Book of the Great Sea-Dragons, Ichthyosauri and Plesiosauri.* William Pickering, London, UK.

Hebda, A.J. 2015. *The Serpent Chronologies: Sea Serpents and other Marine Creatures from Nova Scotia's History – A Book about Stories.* Nova Scotia Museum, Halifax, NS, Canada.

Hendrikx, S. 2018. Monstrosities from the Sea. Taxonomy and Tradition in Conrad Gessner's (1516–1565) Discussion of Cetaceans and Sea-Monsters. *Anthropozoologica* 53:125–137.

Hermannsson, H. 1924. *Jon Gudmundsson and His Natural HIstory of Iceland.* Cornell University Library, Ithaca, NY.

Heuer, C.P. 2019. *Into the White: The Renaissance Arctic and the End of the Image.* Zone Books, Brooklyn, NY.

Heuvelmans, B. 1968. *In the Wake of the Sea-Serpents.* Hill and Wang, New York, NY.

Heuvelmans, B. 1988. The Sources and Method of Cryptozoological Research. *Cryptozoology* 7:1–21.

Higdon, J.W., K.H. Westdal, and S.H. Ferguson. 2014. Distribution and Abundance of Killer Whales (*Orcinus orca*) in Nunavut, Canada—an Inuit Knowledge Survey. *Journal of the Marine Biological Association* 94:1293–1304.

Hill, B., and R. Hill. 1974. *Indian Petroglyphs of the Pacific Northwest.* Hancock House, Surrey, BC, Canada.

Hill, S. 2011. Cryptozoology and Pseudoscience. *Skeptical Briefs* 21:1–3.

Hoage, N. 2017. The Significance of Sea Monsters on Sixteenth Century Maps. Unpublished Master's Thesis. Department of Digital Media Studies, Leiden University, Leiden, Netherlands.

Hoare, P. 2010. *The Whale: In Search of the Giants of the Sea.* Ecco, New York, NY.

Holmes, R. 2008. *The Age of Wonder: How the Romantic Generation Discovered the Beauty and Terror of Science.* Harper Press, New York, NY.

Home, E. 1809. An Anatomical Account of the *Squalus maximus* (of Linnaeus) which in the Structure of its Stomach Forms an Intermediate Link in the Gradation of Animals Between the Whlae Tribe and Cartilaginous Fishes. *Philosoiphical Transactions of the Royal Society of London* 2:206–220.

Houwen, L., and K.E. Olsen. 2001. *Monsters and the Monstrous in Medieval Northwest Europe.* Peeters Publishers, Leuven, Belguim.

Hoyle, W.E. 1902. Sea-Serpent. *Encyclopeida Britannica* [web page]. URL: https://www.1902encyclopedia.com/S/SEA/sea-serpent.html. Accessed on June 3, 2016.

Hunn, E.S. 1978. *Tzeltal Folk Zoology: The Classification of Discontinuities in Nature.* Academic Press, Cambridge, MA.

Hurn, S., ed. 2020. *Anthropology and Cryptozoology: Exploring Encounters with Mysterious Creatures.* Routledge, Abingdon, UK.

Hynes, B. 2012. *Here Be Dragons: Strange Creatures of Newfoundland and Labrador.* Breakwater Books, St. John's, NL, Canada.

Idrobo, C.J., and F. Berkes. 2012. Pangnirtung Inuit and the Greenland Shark: Co-Producing Knowledge of a Little Discussed Species. *Human Ecology* 40:405–414.

Jaffe, A. 2013. Sea Monsters in Antiquity: A Classical and Zoological Investigation. *Berkeley Undergraduate Journal of Classics* 1:1–12.

Jaffe, M. 2000. *The Gilded Dinosaur: The Fossil War between E.D. Cope and O.C. Marsh and the Rise of American Science.* Crown, New York, NY.

Jensz, C. 2012. The Publication and Reception of David Cranz's 1767 *History of Greenland. The Library* 13:457–472.

Johnson, T. 2005. *Entanglements: The Intertwined Fates of Whales and Fishermen.* University Press of Florida, Gainesville, FL.

Johnston, E.M., L.G. Halsey, N.L. Payne, A.A. Kock, G. Iosilevskii, B. Whelan, and J.D.R. Houghton. 2018. Latent Power of Basking Sharks Revealed by Exceptional Breaching Events. *Biology Letters* 14:1–4.

Jones, D. 1989. Doctor Koch and his 'Immense Antediluvian Monsters.' *Alabama Heritage* 12:1–12.

Jorgensen, D. 2018. Beastly Belonging in the Premodern North. In *Visions of North in Premodern Europe*, edited by D. Langum and V. Langum, pp. 183–205. Brepols Publishers, Turnhout, Belgium.

Jylkka, K. 2018. 'Witness the Plesiosaurus': Geological Traces and the Loch Ness Monster Narrative. *Configurations* 26:207–234.

Kapel, F.O. 2005. *Otto Fabricius and the Seals of Greenland.* Meddelelser om Grønland, Bioscience vol 55. Museum Tusculanum Press, Copenhagen, Denmark.

Kendrick, M., and M. Manseau. 2008. Representing Traditional Knowledge: Resource Management and Inuit Knowledge of Barren-Ground Caribou. *Society and Natural Resource* 21:404–418

Kittinger, J.N., L. McClenachan, K.B Gedan, and L.K. Blight, eds. 2015. *Marine Historical Ecology in Conservation: Applying the Past to Manage for the Future.* University of California Press, Berkeley, CA.

Knoepflmacher, U.C., and G.B. Tennyson, eds. 1977. *Nature and the Victorian Imagination.* University of California Press, Berkeley, CA.

Kuban, G.J. 1997. Sea-Monster or Shark? An Analysis of a Supposed Plesiosaur Carcass Netted in 1977. *Reports of the National Center for Science Education* 17:16–28.

Laist, D.W. 1997. Impacts of Marine Debris: Entanglement of Marine Life in Marine Debris Including a Comprehensive List of Species with Entanglement and Ingestion Records. In *Marine Debris*, edited by J.M. Coe and D.B. Rogers, pp. 99–139. Springer, New York, NY.

Landrin, A. 1875. *The Monsters of the Deep, and Curiosities of Ocean Life: A Book of Anecdotes, Traditions, and Legends.* T. Nelson and Sons, Nashville, TN.

LeBlond, P.H., and E.L. Bousfield. 1995. *Cadborosaurus, Survivor from the Deep*. Horsdal & Schubart, Victoria, BC, Canada.

Lee, H. 1883. *Sea Monsters Unmasked*. William Clowes and Sons, London, UK.

Lehn, W.H. 1979. Atmospheric Refraction and Lake Monsters. *Science* 205:183–185.

Lehn, W.H., and I.I. Schroeder. 1981. The Norse Merman as an Optical Phenomenon. *Nature* 289:362–366.

Lehn, W.H., and I.I. Schroeder. 2003. *Hafgerdingar*: A Mystery from the *King's Mirror* Explained. *Polar Record* 39:211–217.

Lehn, W.H., and I.I. Schroeder. 2004. The *Hafstramb* and *Margygr* of the *King's Mirror*: An Analysis. *Polar Record* 40:121–134.

Lenik, E.J. 2010. Mythic Creatures: Serpents, Dragons, and Sea Monsters in Northeastern Rock Art. *Archaeology of Eastern North America* 38:17–37.

Ley, W. 1948. *The Lungfish, the Dodo, & the Unicorn: An Excursion into Romantic Zoology*. The Viking Press, New York, NY.

Lien, J., and L. Fawcett. 1986. Distribution of Basking Sharks *Cetorhinus maximus* Incidentally Caught in Inshore Fishing Gear in Newfoundland. *Canadian Field Naturalist* 100:246–252.

Lilja Bye, B. 2018. Encounters with Giant Sharks in Arctic Waters [web page]. URL: https://blb.as/encounters-with-giant-sharks-in-arctic-waters/. Accessed on June 23, 2019.

Loxton, D., and D.R. Prothero. 2015. *Abominable Science!: Yeti, Nessie, and Other Famous Cryptids*. Columbia University Press, New York, NY.

Lyons, S.L. 2009. *Species, Serpents, Spirits, and Skulls: Science at the Margins in the Victorian Age*. SUNY Press, Albany, NY.

MacNeil, M.A., B.C. McMeans, N.E. Hussey, P. Vecsei, J. Savarsson, K.M. Kovacs, C. Lydersen, M.A. Treble, G.B. Skomal, M. Ramsey, and A.T. Fisk. 2012. Biology of the Greenland Shark *Somniosus microcephalus*. *Journal of Fish Biology* 80:991–1018.

Macrae, J,. and D. Twopeny. 1873. Appearance of an Animal, Believed to Be That Which Is Called the Norwegian Sea-Serpent, on the Western Coast of Scotland, in August, 1872. *The Zoologist* 2(8):3517–3522.

Magin, U. 1996. St George without a Dragon. Bernard Heuvelmans and the Sea Serpent. *Fortean Studies* 4:223–234.

Magin, U. 2016. Necessary Monsters: Claimed 'Crypto-Creatures' Regarded as Genii Loci. *Time & Mind* 9:211–221.

Markham, C. 2015. *The Lands of Silence: A History of Arctic and Antarctic Exploration*. Cambridge University Press, Cambridge, UK.

Martens, F. 1675. Spitzbergische oder Groenlandische Reise-Beschreibung, gethan im Jahre 1671. In *A Collection of Documents on Spitzbergen & Greenland: Comprising a Translation from F. Martens' Voyage to Spitzbergen, a Translation from Isaac de La Peyrère's Histoire du Groenland, and God's Power and Providence in the Preservation of Eight*

Men in Greenland Nine Moneths and Twelve Dayes. No. 18. The Hakluyt Society, London, UK.

Marven, N., and J. James. 2003. Chased by Sea Monsters: Prehistoric Predators of the Deep [Film]. British Broadcasting Corporation, London, UK.

Marven, N., and J. James. 2004. *Chased by Sea Monsters: Prehistoric Predators of the Deep.* DK Publishing, London, UK.

Mason, A. 2017. An Ocean in the Parlor: Home Aquariums were the Newest Fad in the Natural History-Crazed Victorian Era. *Hakai Magazine* [online]. URL: https://www.hakaimagazine.com/features/ocean-parlor/. Accessed on July 4, 2018.

Mathews, L.H., and H.W. Parker. 1950. Notes on the Anatomy and Biology of the Basking Shark *Cetorhinus maximus* (Gunnerus). *Proceedings of the Zoological Society of London* 120:535–576.

Maxwell, G. 1952. *Harpoon at a Venture.* R. Hart-Davis, London, UK.

Mayor, A. 2000. *The First Fossil Hunters: Paleontology in Greek and Roman Times.* Princeton University Press, Princeton, NJ.

Mazzoldi, G. Bearzi, C. Brito, I. Carvalho, E. Desidera, L. Endrizzi, L. Freitas,. G. Giacomello, I. Giovos, P. Guidetti, A. Ressurreição, M. Tull and A. MacDiarmid. 2019. From Sea Monsters to Charismatic Megafauna: Changes in Perception and Use of Large Marine Animals. *PLOS One.* DOI:10.1371/journal.pone.0226810.

McCartney, A.P. 1980. The Nature of Thule Eskimo Whale Use. *Arctic* 33:517–541.

McCaskill, J. 2009. Conserving Waterlogged Rope: A Review of Traditional Methods and Experimental Research with Polyethylene Glycol. Unpublished Master's Thesis, Department of Engineering, Texas A&M University, College Station, TX.

McClenachan, L. 2015. Viewpoint from a Practitioner: The Detective Work of Historical Ecology – New Technologies Enhance, but Cannot Replace Archival Research. In *Marine Historical Ecology in Conservation: Applying the Past to Manage for the Future,* edited by J.N. Kittinger, L. McClenachan, K.B. Gedan, and L.K. Blight, p. 122. University of California Press, Berkeley, CA.

McClenachan, L., F. Ferretti, and J.K. Baum. 2012. From Archives to Conservation: Why Historical Data are Needed to Set Baselines for Marine Animals and Ecosystems. *Conservations Letters* 5:349–359.

McDonnell, M.J., and S.T.A. Pickett, eds. 1993. *Humans as Components of Ecosystems: The Ecology of Subtle Human Effects and Populated Areas.* Springer, New York, NY.

McGowan-Hartman, J. 2013. Shadow of the Dragon: The Convergence of Myth and Science in Nineteenth Century Paleontological Imagery. *Journal of Social History* 47:47–70.

McGregor, R.K. 1997. *A Wider View of the Universe: Henry Thoreau's Study of Nature.* University of Illinois Press, Champaign, IL.

Medin, D.L,. and S. Atran, eds. 1999. *Folkbiology.* MIT Press, Cambridge, MA.

Meine, C. 1999. It's About Time: Conservation Biology and History. *Conservation Biology* 13:1–3.

Merchant, C. 2003. *Reinventing Eden: The Fate of Nature in Western Culture.* Routledge, Abingdon, UK.

Meurger, M., and C. Gagnon. 1988. *Lake Monster Traditions: A Cross-Cultural Analysis.* Fortean Tomes, London, UK.

Møller, P.R., J.G. Nielsen, S.W. Knudsen, J.Y. Poulsen, K. Sunksen, and O.A. Jørgensen. 2010. *A Checklist of the Fish Fauna of Greenland Waters.* (Zootaxa 2378). Magnolia Press, Auckland, New Zealand.

Mowat, F. 1997. *Sea of Slaughter.* McClelland-Bantam, Toronto, ON, Canada.

Mullis, J. 2019. Cryptofiction! Science Fiction and the Rise of Cryptozoology. In *The Paranormal and Popular Culture: A Postmodern Religious Landscape,* edited by D. Caterine and J.W. Morehead, pp. 240–252. Routledge, Abingdon, UK.

Naish, D. 2012. The Cadborosaurus Wars. *Tetrapod Zoology Blog* [online]. URL: https://blogs.scientificamerican.com/tetrapod-zoology/the-cadborosaurus-wars/. Accessed on January 23, 2019.

Naish, D. 2017. *Hunting Monsters: Cryptozoology and the Reality behind the Myths.* Sirius, London, UK.

Nansen, F. 1911. *In Northern Mists.* AMS Press, New York, NY.

National Oceanic and Atmospheric Administration (NOAA). 2014. *Report on the Entanglement of Marine Species in Marine Debris with an Emphasis on Species in the United States.* National Oceanic and Atmospheric Administration Marine Debris Program, Silver Spring, MD.

Nelson, E.W. 1900. *The Eskimo about Bering Strait.* University of California Libraries, Berkeley, CA.

Newman, E. 1849. The Great Sea-Serpent: An Essay, Showing Its History, Authentic, Fictitious, and Hypothetical. *The Zoologist* 54:1604–1608.

Nicolar, J. 1979 [1893]. *The Life and Traditions of the Red Man.* C.H. Glass & Co. Printers, Bangor, ME. Reprint, Saint Annes Point Press, Ames, IA.

Nielsen, J., R.B. Hedeholm, J. Heinemeier, P.G. Bushnell, J.S. Christiansen, S. Jørgen, J. Olsen, C. Ramsey, R.W. Brill, M. Simon, K.F. Steffensen, and J.F. Steffensen. 2016. Eye Lens Radiocarbon Reveals Centuries of Longevity in the Greenland Shark (*Somniosus microcephalus*). *Science* 353:702–704.

Nigg, J. 2013. *Sea Monsters: A Voyage around the World's Most Beguiling Map.* University of Chicago Press, Chicago, IL.

O'Byrne, D. 2018. The Mosasaurus and Immediacy in Jurassic World. In *Beasts of the Deep: Sea Creatures and Popular Culture,* edited by J. Hackett and S. Harrington, pp. 200–213. John Libbey Publishing, Bloomington, IN.

O'Neill, J.P. 1999. *The Great New England Sea Serpent: An Account of Unknown Creatures Sighted by Many Respectable Persons Between 1638 and the Present Day*. Down East Books, Portland, ME.

Oudemans, A.C. 2007 [1892]. *The Great Sea-Serpent*. Brill, Leiden, BE. Reprint, Coachwhip Publications, Darke County, OH.

Ower, J.B. 2001. Crantz, Martens and the "Slimy Things" in "The Rime of the Ancient Mariner." *Neophilogus* 85:474–484.

Paley, W. 2012 [1802]. *Natural Theology; or, Evidence of the Existence and Attributes of the Deity Collected from the Appearances of Nature*. R. Faulder, London, UK. Reprint, Suzeteo Enterprises, Greenwaood, WI.

Papadopoulos, J.K., and D. Ruscillo. 2002. A Ketos in Early Athens: An Archaeology of Whales and Sea Monsters in the Greek World. *American Journal of Archaeology* 106:187–227.

Parish, H. 2020. "None of Them Could Say They Ever Had Seen Them, but Only Had It from Others": Encounters with Animals in Eighteenth-Century Natural Histories of Greenland. *Animals* 10(11):2024. DOI:10.3390/ani10112024

Parsons, E.C.M. 2004. Sea Monsters and Mermaids in Scottish Folklore: Can these Tales Give Us Information on the Historic Occurrence of Marine Animals in Scotland? *Anthrozoös* 17:73–80.

Pauly, D. 1995. Anecdotes and the Shifting Baseline Syndrome of Fisheries. *Trends in Ecology and Evolution* 10:430.

Paxton, C.G.M. 1998. A Cumulative Species Description Curve for Large Open Water Marine Animals. *Journal of Marine Biological Association United Kingdom* 78:1389–1391.

Paxton, C.G.M. 2009. The Plural of 'Anecdote' Can Be 'Data': Statistical Analysis of Viewing Distances in Reports of Unidentified Large Marine Animals 1758–2000. *Journal of Zoology* 279:381–387.

Paxton, C.G.E., and R. Holland. 2005. Was Steenstrup Right? A New Interpretation of the 16th Century Sea Monk of the Oresund. *Steenstrupia* 29:39–47.

Paxton, C.G.M., E. Knatterud, and S.L. Hedley. 2005. Cetaceans, Sex and Sea Serpents: An Analysis of the Egede Accounts of a 'Most Dreadful Monster' Seen Off the Coast of Greenland in 1734. *Archives of Natural History* 32:1–9.

Paxton, C.G.M., and D. Naish. 2019. Did Nineteenth Century Marine Vertebrate Fossil Discoveries Influence Sea Serpent Reports? *Earth Sciences History* 38:16–27.

Paxton, C.G.M., and A.J. Shine. 2016. Consistency in Eyewitness Reports of Aquatic 'Monsters'. *Journal of Scientific Exploration* 30:16–26.

Penn, Z., and W. Herzog. 2004. Incident at Loch Ness [Film]. 20th Century Fox, Los Angeles, CA.

Perdikaris, S., and T.H. McGovern. 2008. Codfish and Kings, Seals and Subsistence: Norse Marine Resource Use in the North Atlantic. In *Human Impacts on Ancient Marine*

Ecosystems: A Global Perspective, edited by T.C. Rick and J.M. Erlandson, pp. 187–214. University of California Press, Berkeley, CA.

Perdikaris, S., and T.H. McGovern. 2009. Viking Age Economics and the Origins of Commercial Cod Fisheries in the North Atlantic. In *Beyond the Catch: Fisheries of the North Atlantic, the North Sea and the Baltic, 900–1850,* edited by L. Sicking, and D. Abreu-Ferreira, pp. 61–90. Brill, Leiden, Belguim.

Perry, S. 2016. *The Essex Serpent.* Serpent's Tail, London, UK.

Peters, R.H. 1980. From Natural History to Ecology. *Perspectives in Biology and Medicine* 23:191–203.

Petersen, H.C. 2000. The Norse Legacy in Greenland. In *Vikings: The North Atlantic Saga,* edited by W.W. Fitzhugh and E.I. Ward, pp. 340–349. Smithsonian Institution Press, London, UK.

Pontoppidan, E.L. 1753. *Det Forste Forsog paa Norges Naturlige Historie.* Lille, France.

Pontoppidan, E.L. 1755. *The Natural History of Norway.* A. Linde, Wien, Austria.

Posey, D.A. 2000. *Cultural and Spiritual Values of Biodiversity.* Practical Advances Publications, New York, NY.

Post-Lauria, S. 1990. "Philosophy in Whales…Poetry in Blubber": Mixed Form in *Moby-Dick. Nineteenth-Century Literature* 45:300–316.

Rea, T. 2001. *Bone Wars: The Excavation and Celebrity of Andrew Carnegie's Dinosaur.* University of Pittsburgh Press, Pittsburgh, PN.

Rees, W.G. 1988a. Polar Mirages. *Polar Record* 24:193–198.

Rees, W.G. 1988b. Reconstruction of an Atmospheric Temperature Profile from a 166-Year Old Polar Mirage. *Polar Record* 24:325–327.

Regal, B. 2011. *Searching for Sasquatch: Crackpots, Eggheads, and Cryptozoology.* Palgrave Macmillian, New York.

Regal, B. 2012. Richard Owen and the Sea-Serpent. *Endeavor* 36:65–68.

Renton, A. 2013. The Basking Shark Returns to British Waters. *The Guardian* [online]. URL: https://www.theguardian.com/environment/2013/oct/13/the-basking-shark-returns-british-waters#:~:text=The%20basking%20sharks%20(or%20the,sea%20was%20colder%20than%20usual. Accessed on March 3, 2019.

Rieppel, L. 2017. Albert Koch's *Hydrarchos* Craze. In *Science Museums in Transition: Cultures of Display in Nineteenth-Century Britain and America,* edited by C. Berkowitz and B. Lightman, pp. 139–161. University of Pittsburgh Press, Pittsburgh, PN.

Rink, H. 1875. *Tales and Traditions of the Eskimo.* AMS Press, New York, NY.

Ritvo, H. 1997. *The Platypus and the Mermaid and Other Figments of the Classifying Imagination.* Harvard University Press, Cambridge, MA.

Ritvo, H. 2010. *Noble Cows and Hybrid Zebras: Essays on Animals and History.* University of Virginia Press, Charlottesville, VA.

Robertson, K. 2020. The Disappearance of Arthur Nestor: Parafiction, Cryptozoology, Curation. *Museum & Society* 18:98–114.

Rosing-Asvid, A., J. Teilmann, R. Dietz, and M.T. Olsen. 2010. First Confirmed Record of Grey Seals in Greenland. *Arctic* 63:471–473.

Rossi, L. 2016. A Review of Cryptozoology: Towards a Scientific Approach to the Study Of 'Hidden Animals.' In *Problematic Wildlife*, edited by F.M. Angelici, pp. 56–76. Springer, New York, NY.

Rotschafer, P.A. 2014. Serpentine Imagery in Nineteenth-Century Prints. Unpublished Master's Thesis, School of Art, Art History and Design, University of Nebraska, Lincoln, NE.

Rudwick, M.J.S. 1992. *Scenes from Deep Time: Early Pictorial Representations of the Prehistoric World*. University of Chicago Press, Chicago, IL.

Rupke, N.A. 1994. *Richard Owen: Victorian Naturalist*. Yale University Press, New Haven, CT.

Saenz-Arroyo, A., C.M. Roberts, J. Torre, and M. Carino-Olvera. 2005. Using Fishers' Anecdotes, Naturalists' Observations and Grey Literature to Reassess Marine Species at Risk: The Case of the Gulf Grouper in the Gulf of California, Mexico. *Fish and Fisheries* 6:121–133.

Saenz-Arroyo, A., C.M. Roberts, J. Torre, M. Carino-Olvera, and J.P. Hawkins. 2006. The Value of Evidence About Past Abundance: Marine Fauna of the Gulf of California Through the Eyes of 16th to 19th Century Travellers. *Fish and Fisheries* 7:128–146.

Sawatzky, H.L., and W.H. Lehn. 1976. The Arctic Mirage and the Early North Atlantic. *Science* 192:1300–1305.

Scheinin, A., D. Kerem, C. MacLeod, M. Gazo, C. Chicote, and M. Castellote. 2011. Gray Whale (*Eschrichtius robustus*) in the Mediterranean Sea: Anomalous Event or Early Sign of Climate-Driven Distribution Change? *Marine Biodiversity Records* 4:10–23.

Schembri, E. 2011. Cryptozoology As a Pseudoscience: Beast in Transition. *Studies by Undergraduate Researchers at Guelph* 5:5–10.

Schowalter, J.E. 1979. When Dinosaurs Return: Children's Fascination with Dinosaurs. *Children Today* 8:2–5.

Scott, P., and R. Rines. 1975. Naming the Loch Ness Monster. *Nature* 258:466–468.

Seaver, K.A. 1996. *The Frozen Echo: Greenland and the Exploration of North America ca A.D. 1000-1500*. Stanford University Press, Redwood City, CA.

Sheets-Pyenson, S. 1981. War and Peace in Natural History Publishing: The Naturalist's Library, 1833–1843. *Isis* 72:50–72.

Sheldon, R.W., and S.R. Kerr. 1972. The Population Density of Monsters in Loch Ness. *Limnology and Oceanography* 17:796–798.

Shermer, M. 2003. Show Me the Body. *Scientific American* 288:37.

Shermer, M. 2010. Hermits and Cranks: Lessons for Martin Gardner on Recognizing Pseu-doscientists. *Scientific American* [online]. URL: https://www.scientificamerican.com/article/hermits-and-cranks-lesson/. Accessed on January 23, 2019.

Shuker, K.P.M. 1996. *In Search of Prehistoric Survivors: Do Giant 'Extinct' Creatures Still Exist?* Blandford Press, London, UK.

Sims, D.W., E.J. Southall, A.J. Richardson, P.C. Reid, and J.D. Metcalfe. 2003. Seasonal Movements and Behavior of Basking Sharks from Archival Tagging: No Evidence of Winter Hibernation. *Marine Ecology Progress Series* 248:187–196.

Skomal, G.B., S.I. Zeeman, J.H. Chisholm, E.L. Summers, H.J. Walsh, K.W. McMahon, W. Kelton, and S.R. Thorrold. 2009. Transequatorial Migrations in Basking Sharks in the Western Atlantic Ocean. *Current Biology* 19:1019–1022.

Soini, W. 2010. *Gloucester's Sea Serpent*. The History Press, Cheltenham, UK.

Solow, A.R., and W.K. Smith. 2005. On Estimating the Number of Species from the Discovery Record. *Proceedings of the Royal Society B* 272:285–287.

Speedie, C. 2017. *A Sea Monster's Tale: In Search of the Basking Shark*. Wild Nature Press, Princeton, NJ.

Starkey, L.J. 2017. Why Sea Monsters Surround the Northern Lands: Olaus Magnus's Conception of Water. *Preternature* 6:31–62.

Stearns, R.P. 1970. *Science in the British Colonies of America*. University of Illinois Press, Champaign, IL.

Steinbeck, J. 1968. *The Log from the Sea of Cortez*. Penguin Classics, New York. Originally published 1951, The Viking Press, New York, NY.

Stothers, R.B. 2004. Ancient Scientific Basis of the "Great Serpent" from Historical Evidence. *Isis* 95:220–238.

Strong, R. 1998. Did Glooscap Kill the Dragon on the Kennebec? *New England Antiquities Research Association Journal* 32:38–43.

Svanberg, I. 1999. The Brother of the Snake and Fish as Kings. *Frooskaparrit* 47:129–138.

Sweeney, J.B. 1972. *A Pictorial History of Sea Monsters and Other Dangerous Marine Life*. Nelson-Crown, New York, NY.

Swinney, G. 1983. The Stronsay Monster: A Case of Mistaken Identity. *Journal of Marine Education* 4:15–17.

Swords, M.D. 1991. The Wago or Sisiutl: A Cryptozoological Sea-Animal of the Pacific Northwest Coast of the Americas. *Journal of Scientific Exploration* 5:85–101.

Szabo, P., and R. Hedl. 2011. Advancing the Integration of History and Ecology for Conservation. *Conservation Biology* 25: 680–687.

Szabo, V.E. 2008. *Monstrous Fishes and the Mead-Dark Sea: Whaling in the Medieval North Atlantic*. Brill, Leiden, Belgium.

Szabo, V. 2018. Northern Seas, Marine Monsters, and Perceptions of the Premodern North Atlantic in the Longue Durée. In *Visions of North in Premodern Europe,* edited by D. Jorgensen, and V. Langum, pp. 145–182. Brepols Publishers, Turnhout, Belgium.

Tatham, D. 1985. Elihu Vedder's Lair of the Sea Serpent. *American Art Journal* 17:33–47.

Taylor, J.G. 1979. Inuit Whaling Technology in Eastern Canada and Greenland. In *Thule Eskimo Culture: An Anthropological Retrospective,* edited by A. McCartney, pp. 292–300. Archaeological Survey of Canada Mercury Paper. No. 88. National Museums of Canada, Ottawa, ON, Canada.

Teit, J.A. 1918. Water-Beings in Shetlandic Folk-Lore, as Remembered by Shetlanders in British Columbia. *Journal of American Folklore* 31:180–201.

Thomas, C. 1988. The 'Monster' Episode in Adomnan's Life of St. Columba. *Cryptozoology* 7:38–45.

Thomas, L. 1996. Appendix: No Super-Otter After All? *Fortean Studies* 3:234–236.

Thomas, L. 2011. *Weird Waters: The Lake and Sea Monsters of Scandinavia and the Baltic States.* CFZ Press, Bideford, UK.

Torrens, H.S. 1991. The Dinosaurs and Dinomania Over 150 Years. In *Vertebrate Fossils and the Evolution of Scientific Concepts,* edited by A.S. Sargent, pp. 255–283. Routledge, Abingdon, UK.

Traill, T.S. 1854. On the Supposed Sea Snake Cast on Shore in the Orkneys in 1808, and the Animal Seen from H.M.S. Daedalus in 1848. *Proceedings of the Royal Society of Edinburgh* 3:208–215.

Van Duzer, C. 2013. *Sea Monsters on Medieval and Renaissance Maps.* The British Library, London, UK.

Van Londen, S. 1996. Mythology and Ecology: A Problematic 'Pas de Deux'. *Cultural Dynamics* 8:25–50.

Vegter, A.C., C. Barletta, J. Beck, H. Borrero, M.L. Burton, M.F. Campbell, M.F. Costa, L.C. Young, and M. Harmann. 2014. Global Research Priorities to Mitigate Plastic Pollution Impacts on Marine Wildlife. *Endangered Species Research* 25:225–247.

Verne, J. 2017 [1870]. *Twenty Thousand Leagues Under the Sea.* Pierre-Jules Hetzel, Paris, France. Reprint, Penguin Classics, New York, NY.

Wabnitz, C., and W.J. Nicols. 2010. Plastic Pollution: An Ocean Emergency. *Marine Turtle Newsletter* 129:1–4.

Wallace, S., and B. Gisborne. 2006. *Basking Sharks: The Slaughter of B.C.'s Gentle Giants.* Transmonanus, Vancouver, BC, Canada.

Walls, L.D. 1995. *Seeing New Worlds: Henry David Thoreau and Nineteenth-Century Natural Science.* University of Wisconsin Press, Madison, WI.

Watanabe, Y.Y., C. Lydersen, A.T. Fisk, and K.M. Kovacs. 2012. The Slowest Fish: Swim Speed and Tail-Beat Frequency of Greenland Sharks. *Journal of Experimental Marine Biology and Ecology* 426: 5–11.

Westrum, R. 1979. Knowledge about Sea Serpents. In *On the Margins of Science: The Social Construction of Rejected Knowledge,* edited by R. Wallis, pp. 234–245. University of Keele, Newcastle-under-Lyme, UK.

Whitaker, I. 1984. Whaling in Classical Iceland. *Polar Record* 22:249–261.

Whitaker, I. 1985. The *King's Mirror* [*Konung's Skuggsja*] and Northern Research. *Polar Record* 22:615–627.

Whitaker, I. 1986. North Atlantic Sea-Creatures in the *King's Mirror* (*Konung's Skuggsja*). *Polar Record* 23:3–13.

White, G. 2016 [1789]. *The Natural History of Selborne.* Self-published. Reprint, Oxford University Press, Oxford, UK.

Whitridge, P. 1999. The Prehistory of Inuit and Yupik Whale Use. *Revista de Arqueología Americana* 16:99–154.

Wikipedia. 2021. Hans Egede [web page]. URL: https://en.wikipedia.org/wiki/Hans_Egede. Accessed on July 2019.

Williams, G. 2015. *A Monstrous Commotion: The Mysteries of Loch Ness.* Orion Books, London, UK.

Wolfson, E. 2001. *Inuit Mythology.* Enslow Publishers, Berkeley Heights, NJ.

Woodley, M.A. 2008. *In the Wake of Bernard Heuvelmans: An Introduction to the History and Future of Sea Serpent Classification.* CFZ Press, Bideford, UK.

Woodley, M.A., C.A. McCormick, and D. Naish. 2012. Response to Bousfield and LeBlond: Shooting Pipefish in a Barrel; or, Sauropterygian 'Mega-Serpents' and Occam's Razor. *Journal of Scientific Exploration* 26:143–147.

Woodley, M.A., D. Naish, and C.A. McCormick. 2011. A Baby Sea-Serpent No More: Reinterpreting Hagelund's Juvenile 'Cadbororsaur' Report. *Journal of Scientific Exploration* 25:497–514.

Woodley, M.A., D. Naish, and H.P. Shanahan. 2008. How Many Extant Pinniped Species Remain to Be Described? *Historical Biology* 20:225–235.

Woods, J.G. 1884. The Trail of the Sea Serpent. *Atlantic Monthly* 53:799–814.

Wyman, J. 1845. A Communication on the Subject of a Fossil Skeleton Recently Exhibited as That of a Sea Serpent. *Proceedings of the Boston Society of Natural History* 2:65–68.

Wyman, J. 1848. 'Hydrarchos sillimani.' *Proceedings of the Boston Society of Natural History* 2:65–68.

Photo Credits

Figure 1.6 (left): File:Hans Egede.jpg. Wikimedia Commons. URL: https://commons.wikimedia.org/wiki/File:Hans_Egede.jpg. Accessed on September 3, 2021.

Figure 1.6 (right): File:Paul egede.png. Wikimedia Commons. URL: https://commons.wikimedia.org/wiki/File:Paul_egede.png. Accessed on September 3, 2021.

Figure 3.6 (middle and lower panels): File:Matham, Jacob - Der am 3. Februar 1598 bei Katwijk gestrandete Potwal - 1598.jpg. Wikimedia Commons. URL: https://commons.wikimedia.org/wiki/File:Matham,_Jacob_-_Der_am_3._Februar_1598_bei_Katwijk_gestrandete_Potwal_-_1598.jpg. Accessed on September 3, 2021.

Figure 4.1 (top-left): File:Gray whale Merrill Gosho NOAA2 crop.jpg. Wikimedia Commons. URL: https://commons.wikimedia.org/wiki/File:Gray_whale_Merrill_Gosho_NOAA2_crop.jpg. Accessed on August 26, 2021.

Figure 4.1 (bottom-left): File:Eschrichtius robustus 01.jpg. Wikimedia Commons. URL: https://commons.wikimedia.org/wiki/File:Eschrichtius_robustus_01.jpg. Accessed on August 26, 2021.

Figure 4.1 (right): File:Anim1723 - Flickr - NOAA Photo Library.jpg. Wikimedia Commons. URL: https://commons.wikimedia.org/wiki/File:Anim1723_-_Flickr_-_NOAA_Photo_Library.jpg. Accessed on August 26, 2021.

Figure 4.5 (top panel): File:Stronsay beast1.jpg. Wikimedia Commons. URL: https://commons.wikimedia.org/wiki/File:Stronsay_beast1.jpg. Accessed on September 3, 2021.

Figure 4.7 (top): Skye Day Five - Basking Shark Breach 2. Photo courtesy of Anthony Robson. URL: https://www.flickr.com/photos/blackpuddinonnabike/4672111013/in/photostream/. Accessed on August 31, 2021.

Figure 4.7 (bottom-left): Still from Science Magazine YouTube video (https://www.youtube.com/watch?v=DPdyU3jY8wQ). Basking shark footage courtesy of Bren Whelan / Donegal

Climbing. URL: http://divemagazine.co.uk/eco/8249-basking-sharks-breaching. Accessed on August 26, 2021.

Figure 4.7 (bottom-right): Skye Day Five - Basking Shark Breach 1. Photo courtesy of Anthony Robson. URL: https://www.flickr.com/photos/blackpuddinonnabike/4672110517/in/photostream/. Accessed on August 31, 2021.

Figure 5.9 (left): Jurassic World trailer: mess with dinosaur DNA at your peril. The Guardian. URL: https://www.theguardian.com/film/filmblog/2014/nov/25/jurassic-world-trailer-dinosaur-dna-chris-pratt. Accessed on September 3, 2021.

www.ingramcontent.com/pod-product-compliance
Lightning Source LLC
Chambersburg PA
CBHW060812270326
41929CB00002B/10

9 780999 075944